Instrumentation and Control

Instrumentation and Control

Sandeep Roy

RANDOM

RANDOM PUBLICATIONS
NEW DELHI (INDIA)

Instrumentation and Control

ISBN 978-93-5111-644-8

Published in 2015 in India by

RANDOM PUBLICATIONS

4376-A/4B, Gali Murari Lal, Ansari Road
New Delhi-110 002
Phone : +9111-43580356, 011-23289044, 011-43142548
e-mail: sales@randompublications.com,
info@randompublications.com, randomexports@gmail.com

Reprinted 2026

Type Setting by : Friends Media, Delhi-110089

Preface

Instrumentation is defined as the art and science of measurement and control of process variables within a production or manufacturing area. The process variables used in industries are Level, Pressure, Temperature, Humidity, Flow, pH, Force, Speed etc. Control engineering or control systems engineering is the engineering discipline that applies control theory to design systems with desired behaviors. Control engineers are responsible for the research, design, development and control devices/systems, typically in manufacturing facilities and plants The practice uses sensors to measure the output performance of the device being controlled and those measurements can be used to give feedback to the input actuators that can make corrections toward desired performance. When a device is designed to perform without the need of human inputs for correction it is called automatic control. Multi-disciplinary in nature, control systems engineering activities focus on implementation of control systems mainly derived by mathematical modeling of systems of a diverse range. This term refers to the graduate discipline which many universities provide at graduate and postgraduate level. Instrumentation and Control plays a significant role in both gathering information from the field and changing the field parameters, and as such are a key part of control loops Instrumentation and control engineers have a role to play in all the fields where there is automation. The instruments created by control engineers to automate the processes, thus reducing the involvement of manpower.

I would like to thank my team for standing beside me throughout my career and writing this book. My special thanks go to "Random Publications" who have published the book.

– Sandeep Roy

Contents

Preface .... *v-vi*

1. Energy Conservation in the Textile Industry **1**

Characteristics of Energy Consumption 1
Promotion of Energy Conservation Technologies 1
Conclusions 17

2. Textile Quality and Cost Control **19**

Management System: An Introduction Definition, Nature, Purpose and Scope of Management 19
Strategies and Management Control 28
Designing Management Control System 38
Pie Chart 77
Run Chart 80
Radar Chart 81
Scatter Plot 83
Histogram 84
Pareto Chart 88
Normality Test 89
Control Chart 91
Normal Test Plot 98
Process Capability 99
Textile Cost Control Techniques 101

3. Functional Suitability of Soluble Peroxidases from Easily control .... in Decolorization of Synthetic Dyes **118**

Peroxidases in Dye Decolorization With Redox Mediators 119
Future Perspective 139

4. Control Methods used in Treatment of Textile Wastewater **140**

Introduction 140

Photocatalysis 141
Photosensitization 142
Advanced Oxidation Processes 145
Fundamental Parameters in Photocatalysis 146
Conclusions 160

5. The Control, Principles and Elements of Textile Design 162

Design Principles 162
The Principle of Digital Printing 164
Design Principles and Elements 166
Graphic or Multimedia Design 171
Inspiration for Textile Designs 177
The Activities of a Textile Designer 179
The Diversity of Textile Design and Textile Designers 188

6. Ironing 196

Introduction 196
Ironing Boards 197
4 Reasons to Love Ironing 200
Steam Iron 202
Ironing-Machine 202

7. Alternative Clothes Cleaning 204

Wet Cleaning 204
Health, Safety and Environmental Regulations 208
Fibres, Yarns, and Fabrics 216
Soils, Odors, Stains and Separation 228
Garment Cleaning Methods 232
Hands-On Wet Cleaning 238
Wet Cleaning Equipment 239

8. Dyeing 242

Introduction 242
Natural Dyeing of Textiles 247
Fabric Dyeing Tips 255

9. Sewing 257

An Overview 257
Sewing Equipment 267

Bibliography 272

Index 274

1

Energy Conservation in the Textile Industry

CHARACTERISTICS OF ENERGY CONSUMPTION

TYPES OF ENERGY USED IN THE TEXTILE INDUSTRY

In general, energy in the textile industry is mostly used in the forms of: electricity, as a common power source for machinery, cooling and temperature control systems, lighting, office equipment, etc.; oil as a fuel for boilers which generate steam; liquified petroleum gas; coal; and city gas. The energy consumption shares of various specialized technical fields and it can be seen that energy consumption is relatively high in the fields of dyeing and finishing, fibre production, spinning, weaving and clothing manufacturing. In the use of various energy sources in the fibre production and dyeing and finishing divisions of the textile industry, where the energy consumption ratio is relatively high.

PROMOTION OF ENERGY CONSERVATION TECHNOLOGIES

While the significance of energy conservation awareness is relatively easily understood at home, when a programme is introduced into a factory to promote it, its thorough implementation tends to be delayed at an early stage. Therefore, for its actual course of implementation, it is desired to devise company-wide coordinated measures similar to QC activities at factories. Also, in order to promote energy saving measures efficiently, it is found to be effective to separately consider general management techniques for "rational use of energy" and process-specific techniques to be developed in each specialized technical field.

ENERGY CONSERVATION MANAGEMENT TECHNOLOGIES

Organizational Rationalization

Since energy management is relevant to a wide range of departments within a company, it is necessary to enhance the awareness, improve the knowledge and obtain the participation and cooperation of everybody involved in the production process. Therefore, while it is necessary for engineers and technicians with specialized technical knowledge to play a central role in energy conservation efforts, the implementation of an energy conservation programme itself should not be left to a handful of specialists or specialized sections. Rather, it is desirable to address the task company-wide, for example by setting up an 'Energy Management Committee'.

Improving Efficiency of Electricity Use

Lighting

Due to its nature of operations, the share of lighting in electricity use is relatively high. After the switch from tungsten bulbs to fluorescent lamps achieved considerable electricity savings, electricity-saving fluorescent lamps have been developed and marketed for further improvements, including those capable of reducing electricity use by several per cent for the same level of illumination. In general, the effectiveness of illumination is influenced by various factors, such as the intensity of light source, the reflection coefficient and shape of the reflective fitting (lamp shade), the layout of the room to be illuminated, interior finish, colour and the distance from the light source. Therefore, it is important to re-examine whether the light source is utilized in the most efficient way and take electricity saving measures, if necessary, such as reducing the number of lamps in use by switching from global lighting to local lighting as much as possible.

Electric motor

The textile industry uses a vast number of relatively small electric motors. Notably, while a conventional machine was driven by a single motor with the generated mechanical power transmitted to various parts of the machine in a collective manner, many modern machines utilize multiple motors with a control board controlling the movement of each motor, which is directly coupled to a machine part to drive it independently from others. This is also a rationalized feature in terms of energy saving. However, regarding the selection of each motor, emphasis has been placed on mechanical performance, resulting in a motor with an excessive capacity. This leaves considerable room for reexamination from a energy conservation point of view.

Electric heating

In the textile industry, electric heating has largely been replaced by other methods (steam, gas heating, or direct or indirect fired heating) for some time in order to achieve cost reductions. However, since electric heating only requires a small initial investment as a result of convenience and simplicity in equipment construction, it is still used for small capacity local heating purposes.

Therefore, it is desirable to conduct a comparative investigation into alternative heating methods, such as far-infrared radiation heating, high frequency dielectric heating and microwave heating.

Improvements in Efficient Fuel Use

Selection of fuel

As is described before, fuels utilized in the textile industry have already gone through a switch-over from coal to oil. More recently, efficient energy use is under investigation, including the revival of coal on the way to a further move from oil to liquefied and city gases, while reflecting various fuel prices. In selecting fuels, those with good flue gas characteristics in addition to high calorific value and ease of combustion are desired, so that air pollution can be prevented as much as possible.

Selection of boiler

By and large, boilers used in the Japanese textile industry have experienced a change from Lancastrian- or Scotch-type tubular or smoke tube to water-tube boilers (natural circulation and forced circulation water-tube boilers and oncethrough boilers). As a result, boiler efficiency has improved from the conventional 60's to 70's of percentage points to as high as the 90's. Since high performance boilers are prone to a rapid growth of scales inside their water tubes, feed water management becomes important. Furthermore, these boilers have small amounts of retained water and high evaporation speeds so that many aspects of their operation are automated, including feed water and combustion management.

Improvement in Efficient Use of Steam

Piping

The noted feature of steam use in the textile industry is that the amount of steam involved is not so large but the locations where steam is required are widespread so that steam losses due to heat radiation from steam transportation pipes and pressure drops are considerable.

Therefore, for steam transportation over long distances, high pressure and smalldiameter rather than low pressure and large-diameter piping is desired, with pressure reducing valves placed as necessary to regulate the steam pressure at the point of use, thereby curbing heat losses.

Also, as pressure losses around bends are great, it is desirable to make their radii large. In order to prevent steam leaks from joints due to the thermal expansion of the pipe, expansion joints should be placed where required. Furthermore, in order to maintain the temperature inside the valve, tank and treatment tank as well as the piping, it is necessary to install them heat-insulated, using appropriate heat insulating materials, so as to efficiently use steam while preventing heat losses.

Steam accumulators

Since live steam is often used in dyeing factories, fluctuations in steam use during working hours are large. On the other hand, since high performance water tube boilers and once-through boilers are designed such that water retained inside the boiler is very little, the boiler cannot react to momentary and sudden load changes, while responding to automatically controlled slow load changes is not a problem.

In such a case, a steam accumulator can be installed midway through the heat transporting pipe, between the boiler and the heat consuming load, in order to store excess steam when the load is light by transforming it to heated water. This then transforms the heated water back to steam when the load is heavy in order to reinforce supply to the load. This allows the boiler to continuously operate with the average load and is quite advantageous in view of energy saving.

Recycling of drain

So far, after its heat energy is consumed, steam has been drained off. However, in view of energy saving, it is necessary to collect and recycle the heat energy carried by the drain water.

Utilization of Heat Exchanger

In each production process of the textile industry, the heating and cooling of gases and liquids as media of heat are frequently required. This is done through heat exchange between different fluids, and in order to avoid contamination or chemical reaction due to their direct contact, heat exchangers are used to carry out indirect heating and cooling. It is important to use the right heat exchanger for the intended purpose.

Measuring Instruments and Automatic Control

Energy saving is an operation to grasp the actual situation of energy use in a factory precisely and quantitatively and to carry out improvement measures in order to rationalize and economize on it.

While measuring instruments are needed to obtain quantitative data, it will become more and more important to investigate the use of sophisticated measuring instruments based on recent developments in mechanical and electronic engineering, combined with automatic control systems.

ENERGY USE AND RATIONAL USE OF ENERGY IN PROCESS-SPECIFIC TECHNOLOGIES

Progress in production rationalization is achieved through the implementation of a comprehensive set of measures, including energy conservation technologies as the centerpiece measure, along with time management, labour saving, natural resources saving and space saving. It has been frequently pointed out that, along with management techniques described earlier, the improvement and development of process-specific techniques on energy conservation greatly contribute to the rationalization of production. Here, process-specific techniques relating to energy saving are summarized for each specialized technical field.

Fibre Production

Exhibiting relatively large-scale structural forms in the textile industry, this division has already reached a high level of production rationalization; as is well known, it is technologically aiming at diversification into such high valueadded goods as super extra-fine fibre and inorganic functional fibre, commonly referred to as *shingosen.* In particular, the following techniques relate to energy saving:

- *Raw material production process:* Implementation of energy saving through improvements in the process and reaction conditions
- *Polymerization process:* Reduction' in polymerization time by means of high efficiency catalysts, polymerization methods, etc.
- *Spinning process:* Promotion of energy saving through combining the POY (Pre-oriented Yarn: Yarn with some stability with its molecules partially having gone through orientation) and DTY (Draw Textured Yam: false twisted yarn produced while drawing POY yarn) methods and an expanded use in multi-folded spinning yarn.
- *Newly built factories:* The factories built during the high growth period have large margins and allowances for production increase so that high losses would result if production decreased. Therefore, suitably sized factories should be constructed.

Spinning

Regarding technological trends in spinning, moves towards high speed and large package size have been investigated in order to achieve labour saving through as much automation as possible. As a result, energy consumption has been gradually increasing. However, in view of price competition with overseas companies, further labour saving as well as energy saving is desired. Table compares a modern and a traditional factory in terms of electricity consumption for each plant/operation.

Namely, it can be seen that a modern factory as a means of achieving production rationalization requires approximately three time as much electricity as a traditional one, with electricity consumption particularly increasing in the air-conditioning plant. In terms of processing operations, fine spinning, as the main operation of the spinning process, consumes a large amount of electricity. Thus, energy saving measures are required in these fields.

Table. Electricity Consumption per 1000 Spindles for each Plant/Process

Factory		Modern Factory		Traditional Factory	
Electricity Consumption		kWh/1000 spindles	%	kWh/1000 spindles	%
Spinning Plant	Mixing/Opening	16.7	7.5	4.2	6.0
	Carding	17.7	8.0	4.9	7.0
	Combing	10.9	5.0	2.2	3.0
	Drawing/Roving	9.1	4.0	2.8	4.0
	Ring Spinning	66.1	30.0	34.5	48.0
	Finishing	14.7	6.5	4.7	6.0
	Subtotal	135.2	61.0	53.3	74.0
Air-conditioning plant		85.9	39.0	18.7	26.0
Total		221.1	100	71.9	100

Ring spinning operation

For the fine spinning operation, electricity is consumed in driving the spindles, packaging, spinning, drafting, and operating the lifting and cleaning mechanisms. It is desired to curb the increase of electricity consumption as much as possible by setting an optimal condition for each of these electricity usages.

Air-conditioning

Although as an ideal working environment a room temperature less than 30°C is desirable, in cases where the working environment has been drastically improved in most other aspects with work load also reduced, a slightly increased room temperature may be permitted. As has been reported, there was a case where raising the regulated temperature from 30°C to 32°C resulted in a reduction in the electric power demand of a carrier with a contract demand of some 8,000 kW by 190 kW. Also, there are many instances of seasonal switch-over from a damper to a pulley as a means of readjusting the blown air volume; this is in order to recycle the air sucked from the processing machine for each operation through a filter back to the same room, and it is therefore necessary to recheck the locations of fans for suction and returning.

Textured-yarn Production

While synthetic-fibre textured-yam is mostly produced with false twisting machines, its history of rationalization is characterized by challenges for high speed operation. As their operating speeds increased, driving and heat-curing motors and and other peripheral equipment became larger, accompanied by an inevitable increase in electricity consumption. Although this may be acceptable as long as the production improvement resulting from a high speed operation covers the increase in electricity costs, reductions in energy cost would surface as an avoidable urgent task, should a sharp increase in electricity charge occur. It can reasonably be said that the major form of energy consumed in the production of synthetic finished-yarn is electricity. Although the amount of electricity consumed in each piece of equipment varies with factory scale and the type of false twist machine, and therefore cannot be treated in a standardized manner, generally accepted average values may be taken as 3.5 kWh/kg for a single heater system and 5.0 kWh/kg for a double heater systemas one report suggests. 60 per cent of all energy consumption by a false twist machine occurs in the heater. Therefore, improvements in the heat insulation of the heater and the lowering of heater temperature may be considered as energy saving measures. Since the latter has implications in the characteristics of the finished-yam, whether or not it is adopted should be examined on such occasions as in the development of a new product. Since air-conditioning plants are designed based on the conditions applicable at the time of installation, it is desirable that they be re-examined against the present conditions.

Weaving

Rationalization in fabric production is such that while various improvements in machinery aimed at high speed operation and labour saving have been carried out, the amount of energy use per unit of the product has gradually increased. Regarding loom design, high productivity shuttleless looms such as water jet, rapier and gripper types have successfully been introduced, with air jet models put in practice in the production area of industrial fabric material. The amount of energy consumed by each loom during its weaving operation can be estimated from the motor capacity and weaving speed. Conventional shuttle looms are based on the weft-insertion method, incorporating a shuttle zooming to and fro with a large inertia mass (approx. 400) and mounted with extra weft, and they also use energy consuming pirns as an integral part of the machine. For this reason, the shuttleless looms' contribution to energy saving cannot be regarded as too high. On the other hand, as a large amount of energy is consumed in sizing, as one of the preparatory operations for weaving, the introduction of foam and solvent sizing operations are being investigated. Furthermore, long fibre fabrics using non-sizing filaments have been developed, eliminating the sizing process altogether. In a reported example, the introduction of a new heat exchanger into a sizing machine with a very poor sealing capability achieved more than 40 per cent of energy saving.

Knitting

The share of energy cost in the total cost of production is not necessarily high for the knitting process. However, of the main production facilities for this process, knitting machines have also been undergoing a shift towards high speed and large capacity and fine gauge features; the current industry trend is for high added-value goods and multi-line, small-volume production based on advanced systems such as computer-controlled pattern making mechanisms. Therefore, a potential tendency for increased energy consumption should be taken into account.

As a result, it is desirable to conduct a comprehensive re-examination of the production schedule along with the implementation of actual energy conservation measures in order to reduce or restrain the share of energy cost in the total production cost.

Dyeing and Finishing

It is very important to advance energy conservation in the dyeing and finishing field, which has a high energy consumption share in terms of both the amounts of money and energy used.

The dyeing and finishing process consists of many interwoven unit operations, and it is well known that the process generally goes through repeated wet and dry operations.

The heat balance of a unit operation can mainly be considered as the difference between the total supplied heat on the one hand and the sum of the heat required by the system and various forms of heat losses on the other.

High speed processing of unit operations

As the processing machines become faster they also become larger. This means the energy consumption per unit length of time will increase, but generally it will accompany a reduction in energy consumption for the treatment of a unit amount of fabric.

Elimination or merger of unit operations

The currently employed dyeing techniques are based on unit operations which have been developed and established for use with natural fibre. For this reason, the traditional standard treatment steps are often applied to blended yam fabrics as a matter of principle.

However, through omitting or merging some of the unit operations according to the usage of the product and considering the characteristics of the coexisting synthetic fibres, it becomes possible to achieve energy conservation.

Reduction in processing bath ratio

It is easy to understand that a reduction in water use will contribute to energy conservation in the dyeing process which consists of various wet treatment and drying unit operations.

It is especially desirable to curb the water consumption because it is linked to the overall water supply cost including that of drainage.

For the reduction of the processing bath ratio, it is necessary to investigate the following measures:

Treatment with low bath ratio

In general, dyeing and finishing methods are classified into the batch and continuous processing methods, and it is recommended to use the latter method where a low bath ratio is desired. However, depending on the details of processing requirements, there are often instances in which the batch method has to be employed. In such cases, batch processing machines which allow lower bath ratios such as the jigger, wince, beam, pad roll and jet flow types should be selected as far as the circumstances permit. Figure graphically shows the relationship between the bath ratio and the production cost for the wince dyeing of a cotton fabric with a reactive dye (bright red, medium shade). It is easy to see that the bath ratio has a direct influence on the production cost,

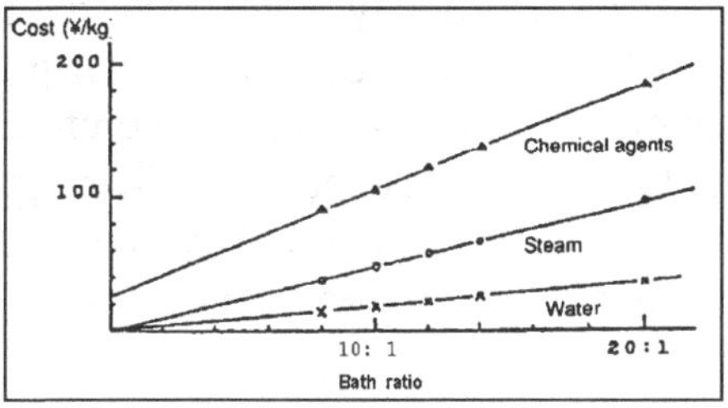

Fig. Relationship between Bath Ratio and Production Cost in Reactive Dyeing

Utilization of low bath ratio processing equipment

In order to use a lower bath ratio with the existing machinery intact, a method to insert a filling material inside the processing equipment has been proposed. It has been reported that with this method, the bath ratio of a wince could decrease from 25:1 to 17:1, and for a beam a reduction was possible from 15:1 to 12.5:1, or even down to as low as less than 10:1 where the axis of the beam was made off center with respect to the container body, thus increasing the batch-up volume as shown in (B). More recently, low bath ratio processing machines which are built in with the above mechanisms have been developed and put on the market.

Utilization of low add-on equipment

Several types of processing equipment with a mechanism to uniformly apply the fabric with a minimum amount of liquid necessary in semicontinuous and continuous processing systems are known to be typical examples of energy conservation techniques.

Extension of foam processing technique

The foam processing technique is used for the preparatory, dyeing, textile printing and finishing processes, with confirmed effects of promoting energy conservation, but it is desirable to examine details of usage and other practical conditions prior to application.

Reduction of processing time

As has been pointed out, being a time saving technique aimed at improving productivity, continuous operation with an increase in the size of the processing machine can also further energy conservation. Likewise, for batch processing, the number of technical fields is increasing where the promotion of energy conservation is desired through a reduction in processing time. This tendency becomes more pronounced as the needs of the market become sophisticated. Techniques to accelerate the processing effect with rapid dyeing and plasma treatment are typical examples.

Rapid Dyeing

Rapid dyeing which can drastically reduce the dyeing time and achieve remarkable time savings can also achieve great energy conservation effects when applied to polyester. In order to attain these effects, it is necessary to select dyes with assistants and provide appropriate dyeing equipment. Combined with the foam processing technique, the rapid dyeing technique may also have a potential of leading up to the development of new practical dyeing techniques.

Accelerating techniques for processing effects

Aiming at a reduction of processing time, the combined use of a number of new techniques are being studied and it has been reported that processing with plasma, ultrasound, magnetism and radioactive rays accelerates processing effects. Various methods are being investigated to reduce processing time through accelerating processing effects using these techniques in preprocessing, postprocessing, simultaneous processing, etc.

Reduction in temperature rise margin

In many cases, unit operations of the dyeing process are carried out at high temperature. Therefore, to reduce the required margin of temperature rise from heating is very important in view of achieving funda-mental energy conservation along with reductions in processing time. These measures need to be addressed from the following two viewpoints:

Raising temperature of inlet water

If the temperature of the inlet water to be used in the dyeing process becomes relatively higher, the amount of energy to be consumed in raising it to the predetermined value will be reduced. For that purpose, cooperation within a company, or that involving more than one company (using low temperature inlet water at the dyeing factory for cooling purposes, and the high temperature discharge from the cooling system for dyeing purposes) should be investigated, as well as the utilization of natural resources (for example, geothermal and solar energy)

Development and introduction of low temperature processing techniques

It is important to continue with technological development aimed specifically at lowering the processing temperature along with raising the inlet water temperature. It would naturally involve the integration of this method with the processing speed acceleration techniques described in (4) (b). Low temperature scouring, bleaching, dyeing and curing techniques are some of the practical examples of this.

Re-examination of drying method

An important consideration along with the reduction of the processing bath ratio is a re-examination of drying operations. A drying operation is, in principle, inserted after every other unit operation and is an important operation which not only determines drying efficiency as such but also has a direct influence on the morphological stability and texture of the final product. For this reason, various types of drying equipment have been selected and put to practical use, depending on the fibre material and form involved. In view of implementing energy saving measures, it is particularly important to investigate the following three items:

Reduction of drying operations in number

A detailed study of typical preparatory process configurations reveals that drying operations are involved in relatively high numbers, ranging from one to four units or 14 per cent to 40 per cent of all unit operations in the entire process. Therefore, combinations of unit operations should be sought after such that drying operations between standard unit operations can be eliminated as much as possible. It is especially necessary to cut down on drying operations in preparatory processes which will not directly affect the product's performance or appearance quality. However, an operation which would have come after an eliminated drying operation would have to treat wet fabrics, thereby necessitating special measures that would enable wet on wet treatment.

Improvements in drying efficiency

It is desirable to investigate possible improvements in the drying efficiency in terms of efficiencies of both the dewatering and drying steps. While it is well known that the most efficient methods of dewatering and drying are by means of a mangle and a cylinder dryer, methods which have been practiced for a long time, they are also known to have limitations in terms of applicable fibre materials and forms. It is necessary to investigate new drying methods (high frequency drying, microwave heating, far infrared radiation heating, etc.) together with other measures such as utilizing vacuum liquid removal, adding a drying-facilitating organic solvent to the treatment liquid, and combining foam treatment systems with non-foam ones.

Recovery of heat energy

Along with active energy saving measures, it is important to carry out the collection and recycling of the energy used in unit operations. The collection of heat energy should start with thoroughly grasping the basic energy balance of each unit operation.

Shift to Solvent Processing

In the dyeing process, although water has been used as the only abundant and cheap resource so far, it is becoming difficult to obtain high quality water in large quantities at a low cost. The worsening of river pollution coinciding with an increase in population density is inevitably creating a situation where the cost of water will gradually increase, including the investment for improvements in waste-water treatment facilities. In addition, although dry-system processing has been contemplated for a long time due to the fact that most energy is consumed in the heating and evaporating operations, it has to date only been applied to a specific area on a limited scale. However, it is a technique which deserves attention as a promising process in the mid- to long-term future. This technique has the following two variations:

Organic solvent processing

While the solvents to be used for dyeing processing are categorized into four main groups—halogenated hydrocarbons, petroleum derivatives, aromatics and oxygen-containing solvents—halogenated hydrocarbons are generally recommended as they do not cause a fire or explosion (provided that thorough countermeasures to groundwater pollution are taken). It is well known that in terms of energy conservation, these solvents have an advantage over water-based ones in all of these aspects: specific heat, latent heat for evaporation, heat needed for evaporation and evaporation speed. There are a number of proposals for solvent scouring, solvent dyeing and solvent finishing, including those already put to practical use as a differentiating technique.

Inorganic solvent processing

Liquid ammonia is one of the agents being considered for dyeing applications as a inorganic solvent. Of its typical processing techniques, liquid ammonia mercerizing and liquid ammonia dying are given particular attention.

Use of Continuous Bath

While textile manufacturing techniques which promote the conservation of natural resources include grease refinement from raw wool, the collection and recycling of warp sizing agents, and the re-use of alkaline waste liquid arising from the mercerizing process in a scouring bath, in terms of energy conservation the use of continuous baths which utilize the residual heat of the system are particularly important in view of energy conservation. If the continuous use of a processing bath is introduced with the necessary conditions being met, thus allowing only those materials consumed in the dyeing process to be replenished, in particular with the unit operations designed for the batch method, it will greatly contribute to the recycling of heat energy in addition to achieving the conservation of natural resources and the rationalization of countermeasures to water waste. In the dyeing process with a high heat consumption, the use of the continuous bath deserves particular attention as a technique whose practical application is an urgent task to help implement the remaining rationalization measures.

Space Saving

With an expansion in the practical use of knit mercerizing and ammonia mercerizing, the characteristics of the hot mercerizing technique is also attracting attention. It has been pointed out that, since poor uniformity associated with the traditional mercerizing due to the hydrophobic nature of cotton grey fabric is dissolved in hot mercerizing which uses heated sodium hydroxide, the rationalization of the preparatory process can be greatly advanced.. At the time a new space-efficient factory is being built, it is possible to incorporate a programme to introduce such facilities as to be complete with factory-wide energy conservation measures.

CLOTHING MANUFACTURING

The energy consumption share of the clothing manufacturing division which consists of large numbers of small-sized companies and their employees in the overall textile industry is not necessarily low but the ratio of energy cost to the total cost is relatively low. However, the energy cost forecast is inevitably a gradual increase under circumstances where the production of high value-added goods is required, along with the implementation of labour saving measures, as a result of the challenging market environment characterized by personalized and diversified consumer needs, high demand for quality goods, short product cycles, etc. Therefore, it is desirable that a comprehensive rationalization programme be investigated apart from reductions in energy consumption.

ACTUAL CONDITIONS OF THE TEXTILE INDUSTRY IN JAPAN

It is widely acknowledged that, as a result of tireless rationalization efforts made by all companies involved, the textile industry has increased its production value by 10 times since 1955 (when it recorded the largest share in total shipment value as a bright star in the entire range of export industries), with a gradual decline in its share over the same period though maintaining a stable growth as a mature industry.

Structural Changes in Textile Industry

It can be seen that large structural changes occurred in the industry as the domestic companies carried out measures to overcome severe competition from overseas, in addition to the already intense competition among themselves. The shares in shipment value of major subdivisions in the textile industry have undergone drastic changes, such as a rapid fall in the fibre production and weaving divisions and a fast growth in the knitting and sewing divisions, reflecting the conditions surrounding the international as well as domestic textile markets.

Progress in Production Rationalization

Despite a strong tendency towards multi-line small-volume production in response to requirements for high value-added products from the fashion market, there is a marked growth in both per capita monthly production volume and product value. Regarding future trends, while the latter is difficult to estimate as it depends on the goods' prices, which are in turn to be determined by a balance between demand and supply, the former may be more easily foreseen as continuing with its increasing trend as a result of advances in the sophistication of production equipment, if profitability is ignored. However, due to the high level of rationalization already incorporated in such equipment, it will be reasonable to expect that attempts to achieve further improvements in this respect will inevitably meet with considerable difficulty.

Changes in Energy Consumption

In the average amount of energy required to make a unit quantity of each textile product over time. It illustrates the course of events as oil replaced coal as the dominant energy source supporting the backbone of society, with the textile industry actively introducing the fluid energy revolution amid high economic growth. This took place against the background of a low level stabilization policy for oil prices made possible by ample oil supplies, including the discovery of large-scale oil fields, as well as oil's convenience in transportation and utilization. It can also be seen that a shift in emphasis from oil with its sky-rocketing price to electricity occurred as a result of energy saving measures introduced after 1973.

More recently, energy saving has been advanced, while a rationalized use of diversified energy sources is being pursued with a global view taking the oil market price etc. into consideration. The oil shock which took place in 1973 encouraged a move away from oil and increased a relative dependence on electricity, comprehensive energy saving measures were also successfully implemented.

Although the achievement of significant energy saving can be observed in the fibre production and dyeing divisions, where energy consumption ratios are particularly high, it is also apparent that their energy consumption has actually been on the increase since 1985. Given this phenomenon along with the steadily increasing energy use in the spinning and weaving fields, it can reasonably be assumed that rationalization efforts are reaching their limits in view of the current production structure designed to cater for the needs for multi-line, small-volume production from the fashion clothing market.

It can easily be seen that the influence of the energy component on the total production cost has been more pronounced after 1973, while at the same time effects of energy saving efforts are also noticeable. The gradual increase in energy consumption in spinning and fabric production and the upward trends after 1985 in fibre production and dyeing are both translated into a decrease in terms of the ratio of energy cost to the total, illustrating that comprehensive energy saving efforts have been made by the companies concerned.

ASSESSMENT OF PRODUCTION RATIONALIZATION LEVEL

Although it is generally difficult to assess the level of production rationalization in the textile manufacturing industry in absolute terms, the dyeing and finishing subdivision is often put under scrutiny as a typical specialized technical field with a number of unknown factors.

It consumes large quantities of energy carrying out multi-line processing tasks within short periods of time in order to directly reflect the market's needs. The results of calculations to find the levels of productivity which could be obtained from dyeing operations assumed to have been conducted in a planned manner at some ideal factories, comparing them with the corresponding average figures for actual Japanese companies.

Since productivity drastically changes with the texture or structure of the original fabric and the details of processing, polyester-cotton blend fabric (shirting of approx 100 g/m2 assumed), which is more likely to be processed in a uniform manner, was used as the material for this scenario.

STRUCTURES OF TEXTILE MARKETS

In general, it is well known that the developmental process of a textile market moves from a product-oriented stage, where all goods produced are sold to a consumer-oriented stage where the desired goods are only those which will satisfy the consumer's demand, comprising his wants and purchasing power. This is via a mass-consumption stage which is supported by a mass-production system. The characteristics of textile products demanded in the markets in these different developmental stages may be classified into price-sensitive mass-production type basic clothing, which applies to the earlier stages, and multi-line, small-volume production type fashion clothing which pertains to the last stage and needs to satisfy each individual consumer's taste.

Basic Clothing

Assuming the amount of fabric necessary for people to live a physically comfortable life in a given environment can be derived from the weight of clothes which is enough to keep a certain level of body temperature, given such factors as air temperature, humidity and wind velocity prevalent in the region, basic clothing can be regarded as a cumulative total of such clothes.

Although the ultimately required quantity is difficult to determine, the per capita annual textile consumption (kg/person) can be estimated through multiplying the above weight by coefficient a (obtained from the service life of a textile product and its number of units demanded per annum).

FASHION CLOTHING

Although it is even more difficult to define the amount of fabric consumed in fashion clothing, a market which has satisfied the demand for minimum basic clothes has a tendency to shift its emphasis towards textile products with stronger fashion overtones and grow rapidly.

WORLD'S TOTAL TEXTILE DEMAND AND PRODUCTION BASE DISTRIBUTION

Assuming a global environment in which world population will grow from the present 5.4 billion to 10 billion in 2050, and further to 11.6 billion in 2150 when it is expected to reach a static state, the total textile consumption is forecast to double, even using the current figure of per capita annual average textile consumption (8kg/person).

The textile industry is traditionally regarded as a typical labour intensive industry developed on the basis of abundant labour supply and has a tendency to expand to overseas markets once the domestic demand is satisfied, as illustrated by examples of the established textile industries of many developed countries.

For this reason, even when the textile industry of a specific country is to be examined, it is widely recognized that a business strategy taking into consideration the global textile industry setup is very important.

CHARACTERISTICS OF TEXTILE MARKET

Most textile consumption takes place in apparel products, and the general tendency is that where further expansion is intended, it will be carried out through the development of industrial applications in which textile products are used as production materials (ranging from fishing nets, tire cords and canvas cloth to geotextile).

Therefore, in order to study the textile manufacturing industry, it is generally sufficient to consider the production of apparel goods.

Specialized Techniques Necessary for Apparel Goods Production

There are a number of intertwined technical factors involved in various subprocesses or specialized technical fields which make up the overall production process that fabricates and reshapes the raw fibre material into the final textile product to be used by the consumer.

Comparison of Characteristics of Specialized Technical Fields

In Fiscal Year 1989, the textile industry had a share of 4.4 per cent in shipment value (19.2 per cent in Fiscal 1955), 14.8 per cent in the number of businesses (20 per cent), and 10.4 per cent in the number of employees (23.3 per cent) in the overall Japanese manufacturing industries, characterized by a gradual decline in relative share, although in terms of absolute shipment value it has actually expanded to 1,053 per cent of the Fiscal 1955 level. As has been frequently pointed out throughout its development, the Japanese textile industry has a unique organizational structure consisting of groups of independent companies where all companies in a group belong to one of the above-mentioned specialized technical fields and operate in a horizontal specialization configuration.

From these indices, common company characteristics for each group may emerge. Namely, in terms of business size, the fibre production and spinning subdivisions are in contrast with the rest of the textile industry where relatively large numbers of small businesses coexist. In addition, these groups of small companies can only stay in business by relying on the supply of abundant cheap labour, exhibiting a legacy of the textile industry's past as a labour intensive industry, even on its path towards modernization.

As for water consumption, the share of the textile industry in the entire manufacturing industries is 5.2 per cent for fresh water and 1.1 per cent for sea water. Considering the fact that most of this water is used for easy-to-recycle temperature control and cooling purposes, the industry's total water consumption cannot necessarily be regarded as high.

However, as is widely accepted, the dyeing and finishing division is placed in a special position in that its water consumption is mainly for processing and washing purposes.

CONCLUSIONS

- There is no panacea for achieving energy conservation in the textile manufacturing industry.
- With the actual implementation of an energy conservation programme, it is important to grasp the current level of energy consumption and its actual conditions in detail, set goals (energy consumption and

corresponding cost), and achieve the goals through a company-wide effort as far as possible.

- In the textile manufacturing industry, it is important to thoroughly understand that, depending on the trend of the market, the company is targeting, consumer requirements for the textile products to be supplied differ, thereby urging the implementation of energy conservation measures which are relevant to the production of the goods that suit the market.
- Therefore, it is necessary to expect that, when multi-line, small-volume production type high value-added goods are produced, energy consumption may increase rather than decrease with production rationalization, in contrast with mass-production type goods.
- When differentiated goods are produced, the share of energy costs in the overall production cost should be given importance rather than energy consumption.
- It is reasonable to consider that ultimately desired energy conservation promoting techniques will depend on the development and practical application of innovative technologies in each specialized technical field.

2

Textile Quality and Cost Control

MANAGEMENT SYSTEM: AN INTRODUCTION DEFINITION, NATURE, PURPOSE AND SCOPE OF MANAGEMENT

AN OVERVIEW

Management is an action procedure composed of some vital functions, for getting the objective of any enterprise accomplished through the efforts of its personnel. Wherever and whenever objectives are to be achieved through organized and co-operative endeavour, management becomes essential for directing and unifying the group efforts towards a general purpose. As human aims and beliefs are mostly realized through the establishment of diverse associations in our civilization, management is universally needed for operating all such organisation.

Management, viewed as a functional concept, is of equal necessity to the educational religious, charitable, and other non-business organizations as it is required for business Organizations. Furthermore, the greatest and the mainly comprehensive of our social organizations, *viz.*, the Government of all kinds require management as others require, perhaps more than all other social organizations. That the Government without requires a management procedure has been applied pointed out in the statement: A Government without good management is a home built on sand. Put in short, management is an essential accompaniment of all social organizations, and it is to be establish everywhere as a separate, separate and control action. The nature and significance of the action do not change even if it is described management in some social in situations and management in others.

Management is the only activating unit of any enterprise for getting things done through it personnel. The job of management is to give dynamic leadership that combines the productive but passive possessions into a fruitful organization. Not only does it adapt itself to existing opportunities, restrictions and pressures, but it exercises a positive power as well as to create the future events favorable for the enterprise. With a view to getting the expected results and seeing that things happen as they should, management has to become a creator of the economy rather then its creature. Management forges ahead through innovations in operating situation and the adoption of distant-sighted scheduling.

It visualizes the future, initiates changes, and achieves the purpose of any enterprise under highly dynamic circumstances. As an action procedure, management plans the future course of action, organizes people and their work, directs the operation and controls the performance, and therefore ensures the accomplishment of enterprise objectives. Adaptations and innovations permeate through each of these phases of the management procedure.

Management acts as a creative and invigorating force in the organization. It makes result that is better than the sum total of efforts put in through the group. Management adds real plus value to the operation of any enterprise through enlisting as small extra value out of each person. It gives new ideas, imaginations and visions to the group working integrate its efforts in such a manner as to explanation for better results. It ensures a smooth flow of work in the organization through focusing on strong points, neutralizing weak link, overcoming difficulties, and establishing team spirit. Management strives to secure the maximum result through the use of minimum possessions.

Management is old as culture and it has been of some concern to organized civilization thought civilized history. Several examples exist today to remind us the result of the organized behaviours of the past. Though, the principles of such organized behaviours are either not accessible or not applicable in present day social structure. In information, no organized effort towards developing management principles was accepted on before the present century and mainly of the earlier contributions towards raising the efficiency of organized clusters were scattered.

Another important point in the context of the development of thought is that the contributions have been made through practicing managers as well as academicians. Therefore the total contribution is the result of practical experience as well as the organized researches. If we examine several thought from era point of view, the role of organized researches successively augment in the development of management principles. For the sake of analysis, the management thoughts have been classified as:

- Systems approach.
- Human dealings approach, and
- Classical Approach

This classification is based on the grouping of several thoughts which have parallel in the approach of how managers get the things done.

SYSTEM APPROACH OF MANAGEMENT

The systems approach, also labeled as contemporary theory of management, is a relatively new approach though neither the conditions management and organisation nor the system is a new concept. In excess of the past tow decades or so the systems approach has appeared as vigorous and lively. Now, we are bombarded with reference to system in several walks of life. Everything is a system and the system thinking is the through work. We have entered an era that forces man to discover more accurate answers to questions involving the 'wholeness' of an operation. This age of synthesis forces management to think out in new and dissimilar ways- as suggested through the systems approach. In this view the whole is not merely a combination of the system but separate from it s parts. Before analyzing how an organisation functions as a system, it is imperative to examine the concept, and working of the system.

A system is defined as the assemblage of things linked or interdependent, so as to form an intricate unity; a whole composed of parts in orderly arrangement just as to some scheme or plan. This has been defined as 'an organized' or intricate whole; an assemblage or combination of things or parts forming an intricate unitary whole.

Several authors on systems approach have attempted to develop a generalized theory which can be applied to any system-physical, biological, or social. They have termed is as Common Systems Theory (GST). The theory is concerned with developing a systematic, theoretical framework for describing the empirical world. Buckley describes its role as such, 'A which functions as a whole through virtue of the inter-dependence of its parts described a system and the method which aims at discovering how this is brought in relation to the in widest diversity of systems has been described Common Systems Theory. Common Systems Theory seeks to classify systems through the way their components are organized and to derive the laws, or typical patterns of behaviour, for the dissimilar classes of systems singled but through the taxonomy.

The first vital level in system management is the determination of organisation system. Since organisation is a deliberate and purposive creation, the vital objectives of the organisation should be determined before its creation. Determination of objectives is significant because every effort is directed towards realization of these objectives. In the second level, each unit of the organisation is arranged in some combination to give desired results. Systems design gives the overall framework for implementing systems concepts. It comprises strategic and comprehensive scheduling for the whole organizational system, as well as the development of operation and facilitating sun-system. The third level-operation and control-refers to conversion of inputs into outputs. Inputs may be in the form of information, materials, and power. The inputs are allocated to plan, though it is possible to eliminate parts of the scheduling required throughout operations through designing system with predetermined input allocation structuring the system to operate in a specified fashion and with more predictable results. Operation of the system requires some sort of control, that is, a mechanism for output or related features, comparing his measurement with the average, and activating the element to adjust inputs to correct the deficiency. This is necessary to uphold the equilibrium of the system close to the ideal point.

The fourth level pertains to review and evaluation, that is, to ascertain how well the system has performed. This is dissimilar from the control in the sense that control refers to operating efficiency, while the concept of review and evaluation is more comprehensive, and always relates the functioning of the system to its objectives. The review and evaluation happen at periodic intervals throughout the life cycle of a system and lead to design changes in the present system or recommendations for changes which may be operated in future systems

Human Relation Approach of Management

The classical approach which focused attention on the mechanical physiological variables of organizational functioning was tested on the field to augment the efficiency of organisations. Surprisingly, positive characteristics of these variables could not evoke positive response in work behaviour, and researchers tried to investigate the reasons for human behaviour at the work. They exposed that the real cause of human behaviour was something more than physiological variables. Such findings generated a new phenomenon in relation to the human behaviour and focused attention of the human beings in the organisations. As such, this new approach his been described 'human approach of management'.

Even in the writings of classical approach, notably, Taylor, Fayol, Henry Gantt, Follet, Urwick, and others, the human unit in the organisation was recognized, but they emphasized it very small. The human dealings approach was born out of a reaction to classical approach and throughout the last four decades, a lot of literature on human dealings had been urbanized. The essence of the human dealings contributions is contained in two points;

- Organizational situation should be viewed in social conditions as well as in economic and technological conditions,
- The social procedure of group behaviour can be understood in conditions of clinical method analogous to the doctor's diagnosis of the human organism.

In the middle of human dealings approach, there are several contributions and several more researches are being accepted on. These contain contributions form well-known 'Hawthorne Experiments', several sociologists-Bakke, Selznic, Homans, dubin, and Dalton; several psychologists-McGregor, Likert, Argyris, March and Simon, Leavitt, Blake, Sayles, Brown, etc. There will be relatively lengthy discussion of the results and implications of the Hawthorne studies because of their historical importance to the behavioural approach to the analysis of management troubles. In information, for the first time, a rigorous and systematic analysis of human factor in organisations was made in the form of Hawthorne experiments.

CONTRIBUTION OF HENRY FAYOL OF CLASSICAL APPROACH

Perhaps, the real father of contemporary management theory is Henry Fayol. He was a French mining engineer, who after obtaining engineering degree joined a French coal firm as technician in 1880. Fayol observed the organizational functioning from manager's point of view.

He establishes that all behaviours of the organisation could be divided into six clusters:

- Technological (relating to manufacture)
- Commercial (buying, selling, and swap)
- Financial (search of capital and its optimum use)
- Accounting (including statistics)
- Managerial (scheduling, organisation, command, co-ordination, and control).

He points out that these behaviours exist in every organisation. He further observes that first five behaviours. *Fayol has divided his approach of learning management in three parts*:

- Managerial qualities and training.
- Common principles of management,
- Units of management.

Management in all business and human organization action is the performance of getting people jointly to accomplish desired goals and objectives. Management comprises scheduling, organizing, staffing, leading or directing, and controlling an organization (a group of one or more people or entities) or effort for the purpose of accomplishing a goal. Resourcing encompasses the deployment and manipulation of human possessions, financial possessions, technological possessions, and natural possessions.

Because organisations can be viewed as systems, management can also be defined as human action, including design, to facilitate the manufacture of useful outcomes from a system.

This view opens the opportunity to 'manage' oneself, a pre-requisite to attempting to manage others. Management can also refer to the person or people who perform the act(s) of management.

The verb *manage* comes from the Italian *maneggiare* (to handle — especially apparatus), which in turn derives from the Latin *manus*. The French word *mesnagement* influenced the development in meaning of the English word *management* in the 17th and 18th centuries. *Some definitions of management are*:

- Organisation and coordination of the behaviours of an enterprise in accordance with sure policies and in attainment of clearly defined objectives. Management is often incorporated as a factor of manufacture beside with machines, materials, and money. The management guru Peter Drucker, the vital task of a management is twofold: marketing and innovation.
- Directors and managers who have the power and responsibility to create decisions to manage an enterprise. As a discipline, management comprises the interlocking functions of formulating corporate policy and organizing, scheduling, controlling, and directing the firm's possessions to achieve the policy's objectives. The size of management can range from one person in a little firm to hundreds or thousands of managers in multinational companies. In big firms the board of directors formulates the policy which is implemented through the chief executive officer.

Mary Parker Follett, who wrote on the topic in the early twentieth century, defined management as "the art of getting things done through people"? She also described management as philosophy.

One can also think of management functionally, as the action of measuring a quantity on a regular foundation and of adjusting some initial plan; or as the actions taken to reach one's planned goal. This applies even in situations where scheduling does not take lay. *From this perspective, Frenchman Henri Fayol considers management to consist of seven functions*:

- Scheduling
- Organizing
- Leading
- Coordinating
- Controlling
- Staffing
- Motivating

Some people, though, discover this definition, while useful, distant too narrow. The phrase "management is what managers do" occurs widely, suggesting the difficulty of defining management, the shifting nature of definitions, and the connection of managerial practices with the subsistence of a managerial cadre or class.

One habit of thought regards management as equivalent to "business management" and therefore excludes management in spaces outside commerce, as for instance in charities and in the public sector. More realistically, though, every organization necessity manages its work, people, procedures, technology, etc. in order to maximize its effectiveness.

Nonetheless, several people refer to university departments which teach management as "business schools."

English speakers may also use the condition "management" or "the management" as a communal word describing the managers of an organization, for instance of a corporation. Historically this use of the condition was often contrasted with the condition "Labour" referring to those being supervised.

In for-profit work, management has as its primary function the satisfaction of a range of stakeholders. This typically involves creation a profit, creating valued products at a reasonable cost (for customers), and providing rewarding employment opportunities (for employees). In non-profit management, add the importance of keeping the faith of donors. In mainly models of management/ governance, shareholders vote for the board of directors, and the board then hires senior management. Some organizations have experimented with other methods (such as employee-voting models) of selecting or reviewing managers; but this occurs only very rarely.

In the public sector of countries constituted as representative democracies, voters elect politicians to public office. Such politicians hire several managers and officers, and in some countries like the United States political appointees lose their jobs on the election of a new president/governor/mayor.

Difficulties arise in tracing the history of management. Some see it (through definition) as a late contemporary (in the sense of late modernity) conceptualization. On those conditions it cannot have a pre-contemporary history, only harbingers (such as stewards).

Others, though, detect management-like-thought back to Sumerian traders and to the builders of the pyramids of ancient Egypt. Slave-owners through the centuries faced the troubles of exploiting/motivating a dependent but sometimes unenthusiastic or recalcitrant workforce, but several pre-industrial enterprises, given their little level, did not feel compelled to face the issues of management systematically.

Given the level of mainly commercial operations and the lack of mechanized record-keeping and recording before the industrial revolution, it made sense for mainly owners of enterprises in those times to carry out management functions through and for themselves.

But with rising size and complexity of organizations, the split flanked by owners (individuals, industrial dynasties, or clusters of shareholders) and day-to-day managers (self-governing specialists in scheduling and control) slowly became more general. While management has been present for millennia, many writers have created a background of works that assisted in contemporary management theories.

Sun Tzu's The Art of War

Written through Chinese common Sun Tzu in the 6th century BC, *The Art of War* is a military strategy book that, for managerial purposes, recommends being aware of and acting on strengths and weaknesses of both a manager's organization and a foe's.

Niccolò Machiavelli's The Prince

Believing that people were motivated through self-interest, Niccolò Machiavelli wrote *The Prince* in 1513 as advice for the leadership of Florence, Italy. Machiavelli recommended that leaders use fear—but not hatred—to uphold control.

Adam Smith's the Wealth of Nations

Written in 1776 through Adam Smith, a Scottish moral philosopher, *The Wealth of Nations* aims for efficient organization of work through Specialization of labour. Smith described how changes in procedures could boost productivity in the manufacture of pins. While individuals could produce 200 pins per day, Smith analyzed the steps involved in manufacture and, with 10 specialists, enabled manufacture of 48,000 pins per day.

19TH CENTURY

Classical economists such as Adam Smith (1723 - 1790) and John Stuart Mill (1806 - 1873) provided a theoretical background to resource-allocation, manufacture, and pricing issues. In relation to the similar time, innovators like Eli Whitney (1765 - 1825), James Watt (1736 - 1819), and Matthew Boulton (1728 - 1809) urbanized units of technological manufacture such as standardization, excellence-control procedures, cost-accounting, interchangeability of parts, and work-scheduling. Several of these characteristics of management lived in the pre-1861 slave-based sector of the US economy. That environment saw 4 million people, as the modern usages had it, "supervised" in profitable quasi-mass manufacture.

Through the late 19th century, marginal economists Alfred Marshall (1842 - 1924), Léon Walras (1834 - 1910), and others introduced a new layer of complexity to the theoretical underpinnings of management. Joseph Wharton offered the first tertiary-stage course in management in 1881.

20TH CENTURY

Through in relation to the 1900 one discovers managers trying to lay their theories on what they regarded as a thoroughly scientific foundation. Examples contain Henry R. Towne's *Science of management* in the 1890s, Frederick Winslow Taylor's *The Principles of Scientific Management*, Frank and Lillian Gilbreth's *Applied motion revise*, and Henry L. Gantt's charts (1910s). In 1912 Yoichi Ueno introduced Taylorism to Japan and became first management consultant of the "Japanese-management approach". His son Ichiro Ueno pioneered Japanese excellence assurance.

The first comprehensive theories of management appeared approximately 1920. The Harvard Business School invented the Master of Business Management degree (MBA) in 1921. People like Henri Fayol (1841 - 1925) and

Alexander Church described the several branches of management and their inter-relationships. In the early 20th century, people like Ordway Tead (1891 - 1973), Walter Scott and J. Mooney applied the principles of psychology to management, while other writers, such as Elton Mayo (1880 - 1949), Mary Parker Follett (1868 - 1933), Chester Barnard (1886 - 1961), Max Weber (1864 - 1920), Rensis Likert (1903 - 1981), and Chris Argyris (1923 -) approached the phenomenon of management from a sociological perspective.

Peter Drucker (1909 – 2005) wrote one of the earliest books on applied management: *Concept of the Corporation* (published in 1946). It resulted from Alfred Sloan (chairman of Common Motors until 1956) commissioning a revise of the organisation. Drucker went on to write 39 books, several in the similar vein.

H. Dodge, Ronald Fisher (1890 - 1962), and Thornton C. Fry introduced statistical techniques into management-studies. In the 1940s, Patrick Blackett combined these statistical theories with microeconomic theory and gave birth to the science of operations research. Operations research, sometimes recognized as "management science" (but separate from Taylor's scientific management), attempts to take a scientific approach to solving management troubles, particularly in the regions of logistics and operations.

Some of the more recent growths contain the Theory of Constraints, management through objectives, reengineering, Six Sigma and several information-technology-driven theories such as agile software development, as well as group management theories such as Cog's Ladder. As the common recognition of managers as a class solidified throughout the 20th century and gave perceived practitioners of the art/science of management a sure amount of prestige, so the way opened for popularized systems of management ideas to peddle their wares. In this context several management fads may have had more to do with pop psychology than with scientific theories of management. *Towards the end of the 20th century, business management came to consist of six separate branches, namely*:

- Human resource management
- Operations management or manufacture management
- Strategic management
- Marketing management
- Financial management
- Information technology management responsible for management information systems

21ST CENTURY

In the 21st century observers discover it increasingly hard to subdivide management into functional categories in this way. More and more procedures simultaneously involve many categories. Instead, one tends to think in conditions of the several procedures, tasks, and objects subject to management.

Branches of management theory also exist relating to non-profits and to government: such as public management, public management, and educational management. Further, management programmes related to civil-civilization organizations have also spawned programmes in non-profit management and social entrepreneurship.

Note that several of the assumptions made through management have approach under attack from business ethics viewpoints, critical management studies, and anti-corporate activism. As one consequence, workplace democracy has become both more general, and more advocated, in some spaces distributing all management functions in the middle of the workers, each of whom takes on a portion of the work. Though, these models predate any current political issue, and may happen more naturally than does a command hierarchy.

All management to some degree embraces democratic principles in that in the extensive condition workers necessity provide majority support to management; otherwise they leave to discover other work, or go on strike. Despite the move towards workplace democracy, command-and-control organization structures remain commonplace and the *de facto* organization structure. Indeed, the entrenched nature of command-and-control can be seen in the way that recent layoffs have been mannered with management ranks affected distant less than employees at the lower stages of organizations. In some cases, management has even rewarded itself with bonuses when lower stage employees have been laid off.

STRATEGIES AND MANAGEMENT CONTROL

CONCEPT OF STRATEGY

The top management of an organization is concerned with the selection of a course of action from in the middle of dissimilar alternatives to meet the organizational objectives. The procedure through which objectives are formulated and achieved is recognized as strategic management and strategy acts as the means to achieve the objective. Strategy is the grand design or an overall 'plan' which an organization chooses in order to move or react towards the set of objectives through by its possessions. Strategies mainly often devote a common programme of action and an implied deployed of emphasis and possessions to attain comprehensive objectives. An organization is measured efficient and operationally effective if it is characterized through coordination flanked by objectives and strategies. There has to be integration of the parts into a complete structure. Strategy helps the organization to meet its uncertain situations with due diligence. Without a strategy, the organization is like a ship without a rudder. It is like a tramp, which has no scrupulous destination to go to. Without an appropriate strategy effectively implemented, the future is always dark and hence, more are the chances of business failure.

Meaning of Strategy

The word 'strategy' has entered in the field of management from the military services where it refers to apply the forces against an enemy to win a war. Originally, the word strategy ha s been derived from Greek, 'strategos' which means generalship. The word as used for the first time in approximately 400 BC. The word strategy means the art of the common to fight in war. The dictionary meaning of strategy is "the art of so moving or disposing the instrument of warfare as to impose upon enemy, the lay time and circumstances for fighting through one self" In management, the concept of strategy is taken in broader conditions. Just as to Glueck, "Strategy is the unified, comprehensive, and integrated plan that relates the strategic advantage of the firm to the challenges of the environment and is intended to ensure that vital objectives of the enterprise are achieved through proper implementation procedure"

This definition of strategy lays stress on the following:

- Unified comprehensive and integrated plan
- Strategic advantage related to challenges of environment
- Proper implementation ensuring attainment of vital objectives

Another definition of strategy is which also relates strategy to its environment. "Strategy is organization's pattern of response to its environment in excess of an era of time to achieve its goals and mission". This definition lays stress on the following:

- It is organization's pattern of response to its environment
- The objective is to achieve its goals and missions

Though, several experts do not agree in relation to the precise scope of strategy. Lack of consensus has lead to two broad categories of definitions: strategy as action inclusive of objective setting and strategy as action exclusive of objective setting.

Strategy as Action, Inclusive of Objective Setting

In 1960's, Chandler made an effort to describe strategy as "the determination of vital extensive condition goals and objective of an enterprise and the adoption of the courses of action and the allocation of possessions necessary for carrying out these goals"

This definition gives for three kinds of actions involved in strategy:

- Determination of extensive condition goals and objectives
- Adoption of courses of action
- Allocation of possessions

Strategy as Action Exclusive of Objective Setting

This is another view in which strategy has been defined. It states that strategy is a way in which the firm, reacting to its environment, deploys its principal possessions and marshalls its efforts in pursuit of its purpose. Michael

Porter has defined strategy as "Creation of a unique and valued location involving a dissimilar set of behaviours. The company that is strategically positioned performs dissimilar behaviours from rivals or performs alike behaviours in dissimilar ways".

The people who consider this version of the definition call strategy a unified, comprehensive, and integrated plan relating to the strategic advantages of the firm to the challenges of the environment After considering both the views, strategy can basically be put as management's plan for achieving its objectives. It basically comprises determination and evaluation of alternative paths to an already recognized mission or objective and eventually, choice of best alternative to be adopted

Nature of Strategy

We can understand the nature of strategy. A few characteristics concerning nature of strategy are as follows:

- Strategy is a major course of action through which an organization relates itself to its environment particularly the external factors to facilitate all actions involved in meeting the objectives of the organization
- Strategy is the blend of internal and external factors. To meet the opportunities and threats provided through the external factors, internal factors are matched with them
- Strategy is the combination of actions aimed to meet a scrupulous condition, to solve sure troubles or to achieve a desirable end. The actions are dissimilar for dissimilar situations
- Due to its dependence on environmental variables, strategy may involve a contradictory action. An organization may take contradictory actions either simultaneously or with a gap of time. For instance, a firm is occupied in closing down of some of its business and at the similar time expanding some
- Strategy is future oriented. Strategic actions are required for new situations which have not arisen before in the past
- Strategy requires some systems and norms for its efficient adoption in any organization
- Strategy gives overall framework for guiding enterprise thinking and action

The purpose of strategy is to determine and communicate a picture of enterprise through a system of major objectives and policies. Strategy is concerned with a unified direction and efficient allocation of an organization's possessions.

A well made strategy guides managerial action and thought. It gives an integrated approach for the organization and aids in meeting the challenges posed through environment

Essence of Strategy

Strategy, just as to a survey mannered in 1974, comprises the determination and evaluation of alternative paths to an already recognized mission or objective and eventually, choice of the alternative to be adopted. Strategy is characterized through four significant characteristics:

- Extensive condition objectives
- Competitive Advantage
- Vector
- Synergy

STRATEGIES AND CORE COMPETENCIES

The Core Competence of the Corporation, C.K. Prahalad and Gary Hamel coined the condition core competencies, or the communal learning and coordination skills behind the firm's product rows. They made the case that core competencies are the source of competitive advantage and enable the firm to introduce an array of new products and services. Just as to Prahalad and Hamel, core competencies lead to the development of core products. Core products are not directly sold to end users; rather, they are used to build a superior number of end-user products. For instance, motors are a core product that can be used in wide array of end products. The business elements of the corporation each tap into the relatively few core products to develop a superior number of end user products based on the core product technology. This flow from core competencies to end products is shown in the following diagram:

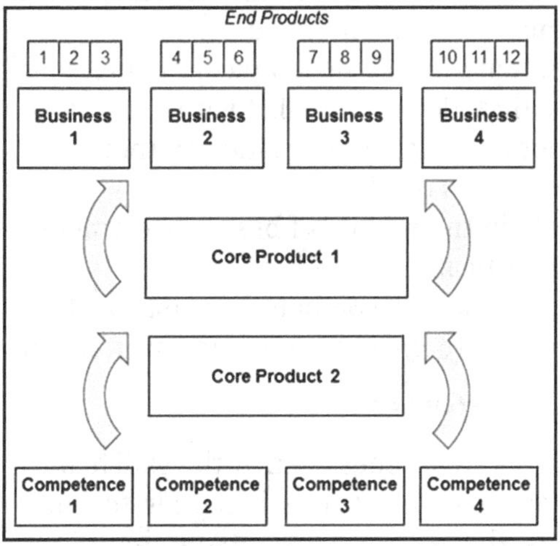

Fig. Core Competencies to End Products

The intersection of market opportunities with core competencies shapes the foundation for launching new businesses. Through combining a set of core competencies in dissimilar ways and matching them to market opportunities, a corporation can launch a vast array of businesses. Without core competencies, a big corporation is presently a collection of discrete businesses. Core competencies serve as the glue that bonds the business elements jointly into a coherent portfolio.

Developing Core Competencies

Just as to Prahalad and Hamel, core competencies arise from the integration of multiple technologies and the coordination of diverse manufacture skills. Some examples contain Philip's expertise in optical media and Sony's skill to miniaturize electronics.

There are three tests useful for identifying a core competence. A core competence should:

- Give access to a wide diversity of markets, and
- Contribute significantly to the end-product benefits, and
- Be hard for competitors to imitate.

Core competencies tend to be rooted in the skill to integrate and coordinate several clusters in the organization. While a company may be able to hire a team of brilliant scientists in a scrupulous technology, in doing so it does not automatically gain a core competence in that technology. It is the effective coordination in the middle of all the clusters involved in bringing a product to market those results in a core competence. It is not necessarily an expensive undertaking to develop core competencies. The missing pieces of a core competency often can be acquired at a low cost through alliances and licensing agreements. In several cases an organizational design that facilitates sharing of competencies can result in much more effective utilization of those competencies for small or no additional cost.

To better understand how to develop core competencies, it is worthwhile to understand what they do not entail. Just as to Prahalad and Hamel, core competencies are not necessarily in relation to the:

- Outspending rivals on R&D
- Sharing costs in the middle of business elements
- Integrating vertically

While the structure of core competencies may be facilitated through some of these actions, through themselves they are insufficient.

The Loss of Core Competencies

Cost-cutting moves sometimes destroy the skill to build core competencies. For instance, decentralization creates it harder to build core competencies because autonomous clusters rely on outsourcing of critical tasks, and this outsourcing prevents the firm from developing core competencies in those tasks since it no longer consolidates the know-how that is spread throughout the company. Failure to recognize core competencies may lead to decisions that result in their loss. For instance, in the 1970's several U.S. manufacturers divested themselves of their television manufacturing businesses, reasoning that the industry was mature and that high excellence, low cost models were accessible from Distant East manufacturers. In the procedure, they lost their core competence in video, and this loss resulted in a handicap in the newer digital television industry. Likewise, Motorola divested itself of its semiconductor DRAM business at 256Kb stage, and then was unable to enter the 1Mb market on its own. Through recognizing its core competencies and understanding the time required building them or regaining them, a company can create better divestment decisions.

Core Products

Core competencies manifest themselves in core products that serve as a link flanked by the competencies and end products. Core products enable value creation in the end products. Examples of firms and some of their core products contain:

- 3M - substrates, coatings, and adhesives
- Black and Decker - little electric motors
- Canon - laser printer subsystems
- Matsushita - VCR subsystems, compressors
- NEC - semiconductors
- Honda - gasoline powered engines

The core products are used to launch a diversity of end products. For instance, Honda uses its engines in automobiles, motorcycles, lawn mowers, and portable generators. Because firms may sell their core products to other firms that use them as the foundation for end user products, traditional events of brand market share are insufficient for evaluating the success of core competencies. Prahalad and Hamel suggest that *core product share* is the appropriate metric. While a company may have a low brand share, it may have high core product share and it is this share that is significant from a core competency standpoint. Once a firm has successful core products, it can expand the number of uses in order to gain a cost advantage via economies of level and economies of scope.

Implications for Corporate Management

Prahalad and Hamel suggest that a corporation should be organized into a portfolio of core competencies rather than a portfolio of self-governing business elements. Business element managers tend to focus on getting immediate end-products to market rapidly and usually do not feel responsible for developing company-wide core competencies. Consequently, without the incentive and direction from corporate management to do otherwise, strategic business elements are inclined to under invest in the structure of core competencies. If a business element does manage to develop its own core competencies in excess of time, due to its autonomy it may not share them with other business elements. As a solution to this problem, Prahalad and Hamel suggest that corporate managers should have the skill to allocate not only cash but also core competencies in the middle of business elements. Business elements that lose key employees for the sake of a corporate core competency should be recognized for their contribution.

CORPORATE STAGE STRATEGY

Corporate stage strategy covers the strategic scope of the organization as a whole. For mainly organizations the corporate strategic plan is the only strategic plan required. Often strategy at the corporate stage is basically referred to as corporate strategy, or in unified companies the corporate business strategy. The procedure that produces it is described corporate strategic scheduling, or sometimes basically corporate scheduling. In a few situations though, it may be justified to speak of corporate stage strategy to distinguish it from other types of scheduling.

Strategy Stages

In the first case the organisation may be multidivisional in nature to the extent that in principle or even in law, separate parts of the enterprise could operate as viable entities in their own right. These 'group structures' may undertake strategic scheduling as group exercise where under the corporate stage strategy, each separate subsidiary or division has its own strategic scheduling procedure and strategic plan. In these cases though, one of the mainly important inputs to each divisions' strategic scheduling is the output of the corporate strategic scheduling.

These outputs from corporate stage strategy; usually in the form of performance targets for the divisions cannot be ignored through the subsidiary element. The corporate business strategy may also set down a little number of other factors that the divisions, or strategic business elements as they may sometimes be described. These might contain guidance on market definition, including geographic scope. For instance the subsidiaries of a multinational bank may be defined through the country they operate in.

In this case the corporate business strategy would set profit targets for each country bank.

The corporate strategy would yield to the country banks as to the strategies they pursue in generating these profits. The country stage banks would have their own business element stage strategies. In the second case corporate stage strategy is used to distinguish it from the several other plans and scheduling procedures that get the condition 'strategic' in their names.

The word strategy has acquired a type of aura that appears to create several people want to use it, regardless of how actually strategic the matter at hand is in relation to the overall performance of an organisation. So we can end up with strategic plans for every stage, part and functional procedure in the organization.

Strategic scheduling is a systematic, formally documented procedure for deciding the handful of key decisions that an organisation, viewed as a corporate whole, necessity get right in order to thrive in excess of the after that few years. Though, because of this wide spread usage in a diversity of contexts we also use the account 'corporate stage strategy' or 'corporate strategy', and refer sometimes to 'corporate strategic scheduling' to create it clear we are not talking in relation to the all these other partial or 'non- corporate' shapes. Because the successful implementation of corporate stage strategy relies on cooperation and alignment crossways the organization as a whole, it is useful to distinguish the several stages of strategy.

Decisions in Corporate Stage Strategy

Keep in mind that at the beginning we said that corporate-stage strategies address the whole strategic scope of the enterprise. This is the "big picture" view of the organization and may contain deciding in which product or service markets to compete and the geographic boundaries of the organizations' operations. For multi-divisional organizations or enterprises, how capital, staffing, and other possessions are allocated is usually recognized at the corporate stage. Additionally, because market definition is usually the domain of corporate-stage strategy, the responsibility for diversification, or the addition of new products or services to the existing offerings, also mostly comes within the responsibility of corporate-stage strategy. Also, whether to compete head on with other companies or to selectively set up cooperative partnering arrangements, or 'strategic alliances' is a decision for corporate-stage strategy, while requiring ongoing input from business element or divisional stage managers.

Corporate Stage Strategic Questions

So crucial questions addressed through corporate-stage strategy, in the middle of other possibilities may contain:

- What should be the scope of operations; *i.e.*; what businesses should the firm be in? And where should it be in business?
- How should the organization allocate its possessions its several existing rows of business or business elements?
- What stage of diversity should exist in the business as it moves into the future? Are there other behaviours the enterprise should be in or are there current behaviours that should be targeted for stopped or sold off to others?
- What should be the nature of this diversity or how diversified should the organization be? Should it diversify in alike product or service markets, or into totally dissimilar regions; becoming a more conglomerate entity.

- How should the firm be organized? What will be the boundaries of the enterprise? How will these boundaries impact relationships in the middle of parts of the business, with suppliers, customers and other interest clusters? How will the organizational functions such as product development, manufacture, sharing finance, marketing, sales customer service, etc. fit jointly? Are the responsibilities for each business element clearly recognized and is accountability recognized? Which will be accepted out in-home, and which will be contracted out?
- Should the firm enter into cooperative, mutually-beneficial relationships or alliances with others? If so, on what foundation? If not, what impact might this have on future organizational performance?

As these questions illustrate, corporate strategies address the extensive-condition direction for the organization as a whole. Corporate strategies deal with plans for the whole organization and change as the capabilities of the organization develop and as the environment of the organization changes. Top management has primary decision creation responsibility in developing corporate strategies and these managers are directly responsible to providers of capital to the organization, whether shareholders, donors, members, and so on depending on the kind of organization. The role of the governing board of is to ensure that top managers actually act to address these owner or primary beneficiary interests.

Business-Stage Strategies

Business-stage strategies are alike to corporate-strategies in that they focus on overall performance. As separate from corporate-stage strategy, though, they focus on presently one instead of a range of businesses. The corporate stage strategy of a multi division operation is like a strategy for managing an investment portfolio. Business elements are usually individual enterprise-like entities oriented towards a scrupulous industry, product or service kind, and or market. Business-stage strategies are therefore primarily concerned with:

- Managing element behaviours so they conform to organizational corporate stage strategies, sometimes including cooperation with other business elements to achieve 'strategic synergy'.
- Developing distinctive capabilities, possessions and competitive advantage in each element.
- Identifying product or service-market opportunities and developing strategies for succeeding in each.
- Monitoring the business industry environment so that strategies conform to requires of the markets at the current level of development.

In a single-product company, corporate-stage and business-stage strategies are the similar. Business-stage strategies seem at the business element strengths, weaknesses, opportunities and threats; much like corporate-stage strategies, except for the emphasis in business-stage strategies is on the specific product or service, not on the corporate stage investment portfolio. Business-stage strategies therefore contribute to corporate-stage strategies. Corporate-stage strategies effort to deliver benefits to the primary beneficiaries, such as raising the wealth of shareholders through profitability of the overall corporate portfolio, and business-stage strategies are concerned wit:

- Matching their operations with the overall objectives of corporate-stage strategy while simultaneously
- Navigating the environment in which they are active in such a way that they are in the middle of the better performers in their industry.

Functional-Stage Strategies

Functional-stage strategies are concerned with managing the functional regions of the organization, such as product or service development and design, marketing and sales, finance, human possessions, manufacture, research and development, etc., so that each function upholds contributes to individual business element strategies and the overall corporate-stage strategy. Functional strategies are primarily concerned with:

- Efficiently deploying specialists within the functional region.
- Integrating behaviours within the functional region
- Creation sure that functional strategies link effectively and efficiently with business strategies and the overall corporate-stage strategy.

BUSINESS ELEMENT STRATEGIES

In an effort to create the traditional business model less bureaucratic and more flexible, some organizations have begun implementing business element strategy.

Instead of forcing a new department into the average chain of command, companies form differently sized autonomous business elements that statement directly to top management. Now the role of the corporation is to manage its business elements and coordinate their efforts into the overall corporate strategy.

Elements

A business element is a part or department of a business that runs as an autonomous entity. Their profits are usually treated separately than those of the business as a whole. "All SBUs (little business elements) are a single business (or collection of businesses), have their own competitors and a manager accountable for operations, and can be independently planned for," reports Reference for Business.

Objectives

Each business element necessity meets five criteria. First, their mission and focus necessity be dissimilar from all other organization SBUs. Second, they should have an easily and clearly defined group of competitors. Third, its scheduling and development should be accepted out quite independently of any other business unity. Finally, each SBU necessity has a separate manager with decision-creation power who is fully responsible for the operation.

Approach

While there is no generic approach that works best all the time, once you determine an approach that works, it's best to stick with it. A collaborative approach builds business elements through relying on employee input combined with corporate strategy. A corporate led approach relies on the executives to plan the element and the goals focus on growth, profit and market share. Thirdly, the business element may form its own strategic plan that is approved, edited or disapproved through corporate management.

Analysis

Business element strategy relies on answering two questions: what are an element's strengths and what is the best way to improve performance. This analysis allows you to set up "the SBU's mission, setting objectives, and determining strategy to use to meet these objectives," reports MAB Paradigms Consulting Firm. SBUs require constant analysis through reports and financial projections to determine if they remain productive or if their strategy necessitates alteration to better improve performance.

DESIGNING MANAGEMENT CONTROL SYSTEM

The three purposes of this level are to give a review of empirical, contingency-based research as it has urbanized since the early 1980s, to critically evaluate this work, and consider a diversity of theoretical foundations that may assist in developing future research. The review is based, in the main, on research employing survey-based methods that has been published in a broad selection of accounting and management journals. The review is selective and illustrative of issues pertinent to the development of a contingency-based framework for the design of MCS, and does not effort a comprehensive coverage of relevant research The chapter is structured as follows. The after that part introduces the region of contingency-based MCS research and gives an overview of findings in excess of the past 20 years. The following nine parts review articles in conditions of their contribution to understanding topics measured within contingency-based research. These are: the meaning of MCS, outcomes of MCS, and the contextual variables of external environment, technology (traditional and modern), organizational structure, size, strategy and national civilization.

Each part comprises two parts: first, findings from the extant literature are presented and a series of propositions summarizing these findings are offered; and second, critical issues concerning each variable are examined with a view to identifying regions that give challenges for improvement and opportunities for future research. Following these parts, issues concerning theory development are examined. Finally, the potential role of a diversity of theories in progressing understanding of contingency-based research in MCS is measured.

AN ORGANIZATIONAL FRAMEWORK FOR CONTINGENCY BASED MCS RESEARCH

The identification of contextual variables potentially implicated in the design of effective MCS can be traced to the original structural contingency frameworks urbanized within organizational theory. Theorists such as Burns and Stalker, Perrow, Thompson, Lawrence and Lorsch, and Galbraith focused on the impact of environment and technology on organizational structure. Early accounting researchers drew on this work to investigate the importance of environment, technology, structure and size to the design of MCS. Reviews mannered 20 years ago through Waterhouse and Tiessen and Otley were able to structure their commentaries through categorizing the early research into these key variables.

In considering MCS research since 1980, it is apparent that these key variables have been confirmed as descriptors of fundamental, generic units of context. Several recent studies, incorporated in this review, focus on modern characteristics of the environment, technologies and structural arrangements. They draw on the original organizational theorists to develop arguments that help explain how the effectiveness of MCS depends on the nature of modern settings.

Also, recent research has measured the relevance of additional contextual variables to the design of MCS. Perhaps the mainly significant new stream of literature has been that related to the role of strategy. This has been assimilated within the traditional organizational model in ways that suggest significant links flanked by strategy, the environment, technology, organizational structure and MCS. The importance of technology to MCS design has been enriched through research drawing on the manufacturing literature, and the work of economists such as Milgrom and Roberts.

Issues concerning the role of MCS within advanced manufacturing settings such as Total Excellence Management, Presently-in-Time and Flexible Manufacturing have been explored. Researchers have gained new insights into the role of MCS within new structural arrangements, such as teams, through drawing on the human resource management literature which investigates the dynamics of teams including issues concerning performance evaluation. National civilization has been recognized as a unit of context following the development of multi-national operations in several organizations. In reviewing the past 20 years of contingency based research it is significant to consider the extent to which progress has been made in developing an empirical body of literature relating MCS to units of context. The conventional, functionalist contingency-based approach to research assumes that MCS are adopted to assist managers achieve some desired organizational outcomes or organizational goals.

The appropriate design(s) of MCS will be influenced through the context within which they operate. The following nine parts consider: the meaning of MCS, the outcomes of MCS and the key contextual variables as they have evolved, historically, in the literature. First, the connection flanked by MCS and the external environment is measured. This is followed through technology (both traditional and modern), structure and size. After that, strategy is examined. Finally, the role of national civilization in MCS design is reviewed. On the foundation of the empirical findings, propositions are offered which relate contextual variables to the MCS. Assessing these propositions requires considering shortcomings in contingency-based research, identifying the extent to which progress has been made in addressing these issues and noting opportunities for improvements and future directions.

THE MEANING OF MCS

The conditions management accounting (MA), management accounting systems (MAS), management control systems (MCS), and organizational controls (OC)are sometimes used interchangeably. MA refers to a collection of practices such as budgeting or product costing, while MAS refers to the systematic use of MA to achieve some goal. MCS is a broader condition that encompasses MAS and also comprises other controls such as personal or clan controls. OC is sometimes used to refer to controls built into behaviours and procedures such as statistical excellence control, presently-in-time management. The condition MCS is used, in the main, throughout this chapter. The definition of MCS has evolved in excess of the years from one focusing on the provision of more formal, financially quantifiable information to assist managerial decision creation to one that embraces a much broader scope of information.

This comprises external information related to markets, customers, competitors, non-financial information related to manufacture procedures, predictive information and a broad array of decision support mechanisms, and informal personal and social controls. Conventionally, MCS are perceived as passive apparatus providing information to assist managers. Though, approaches following a sociological orientation see MCS as more active, furnishing individuals with power to achieve their own ends. Contingency-based research follows the more conventional view that perceives MCS as a passive tool intended to assist manager's decision creation.

Contingency-based research has focused on a diversity of characteristics of MCS. These contain dimensions of budgeting such as participation, importance of meeting budgets, formality of communications and systems sophistication, links to reward systems, budget slack, post completion audits and variance analysis. Examples of modern innovations in MCS contain ABC/ ABM, non-financial performance events and economic value analysis. Researchers are likely investigating progress concerning balanced scorecards and target costing.

At a more common stage, studies have measured sophistication of controls, reliance on accounting performance events, dimensions of information such as scope, timeliness and aggregations, sophisticated capital budgeting, cost consciousness, competitor focused accounting, strategic interactive controls and diagnostic controls, information which is related to issues concerning customers, product design, time, cost, possessions and profitability which is distinguished on stage of detail, updating frequency and interactive use with operational personnel.

Critical Evaluation

Overall, assessing findings from contingency based research involves judging how the results accumulate to give generalizable findings concerning MCS. As is general in several social sciences, MCS researchers are faced with decisions on whether to build on an existing region of revise, such as the role of formal budgets, or identify emerging characteristics of MCS, such as balanced scorecards or target costing, and investigate the settings within which they may be mainly beneficial. Within the body of literature reviewed in this paper there is a mixture of studies focused on single themes, and studies exploring unconnected units of MCS and context. Both kinds of studies are required. Learning the role of novel MCS practices within modern settings is necessary to ensure that MCS research is relevant. Given that several dimensions of MCS and their contexts change, novel studies will always be required to address emerging issues.

There is a pressing require for studies into situations in which modern MCS may be best suited. Research is presently starting to be published identifying contingencies nearby the design and implementation of ABC/ABM. Though, there is very small published contingency work on balanced scorecards, target costing, life cycle costing, the broad array of non-financial performance indictors including those related to assessing human resource management initiatives. Examples of the latter contain measurement to guide and evaluate the learning capabilities of the organization, events such as team maturity indexes and organizational climate surveys that effort to assess the effectiveness of administrative innovations.

Recent growths in global operations, strategic risk management (including real options), corporate social reporting (including triple bottom row), economic value analysis, 360 degree performance evaluation, knowledge-based organizations, forensic accounting, intellectual capital, competitor accounting and value chain are only presently beginning to be understood through researchers. There is a require for more research into service and not-for-profit organizations as these entities become increasingly significant within mainly economies Notwithstanding the importance of learning controls that are relevant to modern settings, it is significant to develop knowledge in ways that ensure coherence in the revise of units of MAS and contextual variables, and in the findings of these studies. Such confidence can be derived from replication studies which enhance the validity and reliability of findings and thereby give a strong base to move forward through way of model development.

Commentators have been critical that in mainly regions of MCS research, studies have not urbanized enough 'critical mass' to confirm findings. The way in which studies evolved within the region of RAPM helps illustrate many difficulties in isolating the meaning and measurement of MCS variables. First, the precise meaning of the concept of RAPM has been confused through lack of definition of what is accounting and non-accounting and what is reliance. Given the ambiguity with the concept, it is not surprising that researchers sought to gain clarification through modifying their studies as understanding of RAPM and its measurement urbanized.

Such refining of concepts and measurement is general in other social sciences, such as psychology. It is unfortunate that it is not part of the MCS research custom to spend more time on developing robust events of the units of MCS, particularly when there is ambiguity in the meaning of constructs. For instance, it is not clear how balanced scorecards should be measured. It appears likely that the content and implementation of balanced scorecards vary widely flanked by organizations. It would appear useful to develop a valid measure of balanced scorecards that could then be used through researchers to explore its context. While such a valid measure would enhance consistency flanked by studies, a difficulty exists in the dynamic nature of MCS practices.

MCS that are valid today may lose validity as they evolve through time. Certainly, because of advances in information technology (IT)software, some kinds of balanced scorecards being employed today are more comprehensive and strategic in nature than those being used 5 years ago. Likewise, the concept of RAPM and how it relates to broader controls has changed since the early work in the 1970 and 1980s. Without accommodating changes in modern control systems, concepts and events of MCS are unlikely to address pertinent, modern issues.

A research climate that encouraged the development of valid concepts and events of MCS would have to recognize require for modification to incorporate the development of MCS. Participative budgeting has also been studied widely. Unlike RAPM, participation in budgets has approximately universally been conceptualized and measured following Milani. In some studies, additional events are employed to give some validation of the primary measure. Other studies of budget related behaviour have drawn on attitudes and satisfaction with budgets, as urbanized through Swieringa and Moncur.

There have been a considerable number of studies that have confirmed the measurement of the generic MCS features of broad scope, timeliness, aggregation, and integration. These studies have employed concepts and events urbanized through Chenhall and Morris, sometimes with minor adjustments to suit the scrupulous setting, and appear to be robust crossways a diversity of settings. Though, there has been small replication or coherence in measurement development in studies examining MCS practices of modern interest such as static flexible budgets, non-financial performance events, action-based accounting, competitor focused accounting, and product development information. Likewise, while studies have explored significant regions of MCS such as social controls, personnel control, sophisticated integrative mechanisms, administrative controls, interpersonal controls, sophisticated controls, there has been very small replication.

A further criticism related to the nature of accounting controls within contingency-based research is that these form only part of broader control systems. Contingency-based research has focused on specific units of accounting controls, generic information dimensions of MCS, with a limited number of studies examining broader units of control, such as clan and informal controls, or integrative mechanisms. A difficulty in learning specific units of MCS in separation from other organizational controls is the potential for serious model under specification. Therefore, if specific accounting controls are systematically connected with other organizational controls, studies that exclude or do not control for these units within the research method may statement spurious findings. For instance, a revise focused only on formal budget systems may argue that they are unsuitable in uncertain operating circumstances as they contain partial information and lack flexibility.

Though, proof may indicate that successful organizations rely extensively on formal budgets. This unexpected finding occurs as a consequence of limiting the revise to budgets without considering broader control and information networks. It may be that successful organizations operating in uncertain circumstances have formal budgets but they are systematically combined with open and flexible informal communications flanked by managers.

The formal budgets are useful in assisting scheduling and curbing excessive innovation, while the informal communications give broader information in flexible ways. Simons showed that formal budgets can give interactive controls in uncertain circumstances whereby the budgets generate intelligence data to build internal pressure to break out of narrow search routines and encourage the emergence of new strategic initiatives. Chapman also argues that in uncertain circumstances effective organizations can employ formal accounting but they should take lay within a situation that involves intense verbal communication flanked by organizational clusters. A way of addressing these concerns is to identify a diversity of control taxonomies and consider how they relate to several characteristics of MCS. One such taxonomy involves classifying controls as ranging from mechanistic to organic.

Mechanistic controls rely on formal rules, standardized operating procedures and routines. Organic systems are more flexible, responsive, involve fewer rules and standardized procedures and tend to be richer in data. Table gives a grouping of units of MCS and control kinds commonly establish in research, in conditions of the organic or mechanistic nature of control.

These taxonomies are useful for addressing concerns of how MCS relate to broader control systems and can guide research into how scrupulous characteristics of MCS are constant with the control 'civilization' of organizations. Finally, it should be noted that there is a distinction flanked by the adoption of MCS and the implementation of the systems. Much can be learned in relation to the success or otherwise of MCS through examining how the control civilization, organic or mechanistic, powers the procedures of implementation. This becomes particularly significant when learning the adoption of innovative MCS such as action-based accounting and balanced scorecards which are closely connected to the organization's control civilization and the extent of change in MCS, in common. It appears clear that broader issues of control are likely to have implications for research into understanding MCS design. There have been advances in excess of the past 20 years in demonstrating the importance of considering management accounting practices as characteristics of MCS.

Understanding how specific characteristics of management accounting relate to broader control concepts, as outlined in Table, assists in researching the complementary or substitution effects of non-accounting controls. A significant part of the research agenda is to understand how combinations of controls can be combined, to suit the scrupulous circumstances of the organization. In learning broad controls, it is necessary to be aware of the boundaries that some organizations and accountants lay approximately MAS and MCS. Without such awareness, there can be confusion as to what formal accounting control is, what structural control is, and what are personnel and informal controls.

OUTCOMES OF MCS

Outcomes may be separated into issues related to the use or usefulness of the MCS, behavioural and organizational outcomes. There is an implied connection flanked by these outcomes. If the MCS are establish to be useful then they are likely to be used and give satisfaction to individuals, who then presumably can approach their tasks with enhanced information. As a consequence, these individuals take improved decisions and better achieve organizational goals.

Clearly, there are broad leaps in logic from useful MCS, to improved job satisfaction and enhanced organizational performance. Moreover, there is no compelling proof to suggest that such links exist. Even within contingency-based research, the link flanked by enhanced organizational performance and usefulness of some aspect of MCS may well depend on the appropriateness of the useful MCS to the context of the organization. Thoughts of interest to designers and researchers of MCS have been the extent to which the systems give information, the degree of use, the usefulness of the information or the beneficial nature of the MCS; the importance in creation operational decisions; importance to product development, whether they are helpful to the organization, and satisfaction with the systems.

Behavioural outcomes such as job satisfaction have been significant in human resource management. The provision of a work laid environment to enhance employee welfare or job satisfaction is seen through some as a worthwhile goal in its own right. Moreover, other things equal, it may be presumed that individuals who are satisfied with their jobs will identify with organizational goals and work more effectively. Interestingly, there have not been several MCS studies that have examined the effects of MCS on job satisfaction. A diversity of studies have examined the effect of MCS on job related tension or stress. Unlike job satisfaction, stress appears to be more closely related to the nature of the MCS and is implicated in associations with performance.

Organizational outcomes in contingency-based research have been dominated through self assessment procedures where individuals give an indication of their performance, or their organizational element, crossways a range of potentially significant managerial procedures or goals of the organization. The issue of the validity of self-assessment is often raised as a concern. Proof suggests that a subordinate's self-assessment correlates with objective assessments and with a superior's subjective assessment. Notwithstanding this proof, it is always reassuring when a superior's performance rating of the respondent is incorporated in the revise.

Curiously, there has been small work relating MCS change to share prices. Larcker establish that firms adopting incentive performance plans experienced an augment in capital investment and a positive security market reaction on disclosure of the plan to the market. Gordon and Smith accounted that returns to investors were higher for firms employing post completion reviews when matched with asymmetric information, capital intensity, capital expenditure and insider ownership. Smith recognized that positive returns were associated with post completion reviews in abandonment decisions.

Though, Gordon and Silvester establish no important association flanked by the installation of ABC and important stock market reaction. This poses the question as to whether improved understanding would follow from learning these associations within a diversity of organizational contexts. These studies do not employ contingency-based approaches as they explore only the main effects flanked by share price movement and the adoption of units of MCS. Progress in this region may be limited due to the difficulties in extracting the effects of adopting dissimilar MCS on share prices from other events that may be associated with share price movements. With the numerous possible events effecting share prices, control troubles can become acute. Also, data collection is complicated through require to collect data on the adoption and implementation of MCS through survey methods and to then match these with share price changes. Also, perhaps the lack of research in the region says something in relation to the dissimilar kinds of training flanked by researchers in finance and management accounting.

Critical Evaluation

Contingency-based studies have examined MCS as both dependent and self-governing variables. To look at fit flanked by MCS and context, some commentators have claimed that the outcome variables should be some dimension of desired organizational or managerial performance. Good fit means enhanced performance, while poor fit implies diminished performance. While the ultimate goal of MCS research is to give findings that assist managers achieve their goals or those of their organization, MCS research has sustained to contain dimensions of MCS, their use and usefulness, as the outcome variable. Also, it is noteworthy that performance has been incorporated as a self-governing variable explaining some features of MCS. While not explicit in mainly studies with MCS as the outcome variable, it is implied that associations flanked by context and MCS reflect equilibrium circumstances, or indicate optimal solutions because of survival of the fittest circumstances.

If equilibrium is assumed, then learning performance is inappropriate as every firm has optimal performance given its situation. Researchers learning MCS as the outcome variables, also, note that this approach is justified through assuming that rational managers are unlikely to adopt or use MCS that do not assist in enhancing performance. Alternatively, some argue that links flanked by MCS, context and performance can be tenuous as they involve several factors concerning the excellence of managing the manufacture procedures.

In common, if disequilibrium circumstances are assumed, then it may be useful for contingency based studies to first set up adoption and use of MCS, then to look at how they are used to enhance decision excellence and finally investigate links with organizational performance. Care is required when interpreting studies that have outcome variables related to the features of MCS, such as 'use' and 'usefulness' of the systems. Individuals may be forced to use MCS systems, such as budgets or DCF analysis in decision creation, even though they discover them of small use. A scrupulous MCS may be perceived as not useful, rate low in satisfaction or benefits but organizational performance may be high due to the supply of required information from other sources, either formal or informal.

These issues can be resolved through careful attention to the research question. It is quite legitimate to revise the adoption of systems and their use. Though, it may not be appropriate to claim that these outcomes are of value in improving organization performance. Likewise, the extent to which MCS are perceived as useful may not imply improved organizational performance.

If learning one aspect of the MCS in separation from other sources of information, researchers should ensure that the studied attribute is the sole source of the information being studied. If an aspect of the MCS is being measured within situations that contain broader information and controls, the potential power of these other controls should be incorporated or controlled within the research design. In summary, despite the critique that contingency-based studies should contain organizational performance as the dependent variable, studies still follow approaches with MCS as the dependent variable. Care in theory construction is required in following either approach.

Studies can give significant insights into the extent of adoption, use and usefulness of MCS, though, it should not be assumed that the models necessarily lead to enhanced organizational performance. Likewise, if performance is the dependent variable then compelling theory is required to illustrate how the combination of MCS and context enable managers to take more effective decisions that enhance organizational performance. Given the assumption that research should identify organizational performance as the criterion variable, a critical issue is what constitutes organizational performance? Distinguishing official and operative goals would appear an essential aspect of MCS research that comprises consideration of goals, mainly as it flags that the issue of organizational goals is distant from unproblematic.

Investigating these goals requires a dynamic approach that examines the goal formulation procedure. There are many issues that become significant. First, goal formulation or change often involves the power of new powerful players, either within or outside the organization, who can dramatically change official goals. MCS can act either as a tool to effect such changes or hinder their acceptance within the organization. For instance, a new Chief Executive Officer may stipulate that improved shareholder value is a priority.

Consequently, performance measurement based on Economic Value Analysis may be introduced in an effort to align the actions of all employees with the single objective of improving economic value. Second, changes in the regions of organizational effectiveness can redirect goals to those regions of effectiveness. The unplanned detection of a new technology that potentially increases throughput can result in the adoption of 'timely' delivery as a goal of the organization. Third, it is apparent that the measurement of goals can have explicit effects on goal formulation, both planned and unintended. Goals may be selected or evolve as they can be measured readily through the MCS.

A preoccupation with formal, 'hard' events may direct attention to those events at the expense of the subtleties of the situation. For instance, measuring characteristics of customer or employee satisfaction, the organizational civilization or intellectual capital often require more subjective assessments of progress and, as such, may receive less attention than behaviours subjected to hard events such as manufacture rejects or throughput. Fourth, in addition to influencing kinds of goals, MCS may affect goal attainment through establishing standards or benchmarks for performance. Goals that are too hard may cause frustration and withdrawal, while standards that are too easy may not give enough challenge. Standards that are achievable but with enough stretch to give a challenge are often recommended as ideal.

Though, in today's environment of intense competition and global operations, necessities for substantial continuous improvement may mean that hard standards based on continuous improvements are required to survive. Performance events can readily set up targets that require continuous improvement. Fifth, recently several organizations have recognized require to satisfy multiple and potentially competing goals. Mission statements identify the necessities to draw and uphold shareholders, employees, and customers and to do so in ways that are socially acceptable. Accountants have responded through refining triple bottom row reporting, environmental accounting, social corporate reporting and corporate sustainability.

Sixth, aligning operative goals with official goals is a significant aspect of strategic management. This is the essence of performance hierarchies and balanced scorecards that effort to capture the interactive effects of official goals associated with the interests of shareholders, customers, the internal procedures and the potential for the organization to sustain itself through learning and innovation. Of course, connections flanked by official and operative goals can be quite dissimilar in same organizations.

Achieving shareholder welfare might require organizations to follow dissimilar operative goals concerning decisions on excellence, cost, delivery and the like. Linkages flanked by MCS and organizational goals are quite explicit, as a primary function of MCS is to measure progress towards achieving desired organization ends.

It is a useful exercise when evaluating features of MCS used for reporting on goals to judge the extent to which they accommodate the following: consider multiple stakeholders; measure efficiency, effectiveness and equity; capture financial and non-financial outcomes; give vertical links flanked by strategy and operations and horizontal links crossways the value chain; give information on how the organization relates to its external environment and its skill to adapt.

Presumably, balanced scorecards or performance hierarchies give a methodology to tackle several of these issues. The complexities of achieving these expectations may help to explain why several firm that effort to adopt balanced scorecards have difficulty in implementing them.

CONTEXTUAL VARIABLES AND MCS

Before examining the contextual variables, a distinction is noted flanked by generic and specific definitions. When considering environment, specific definitions refer to scrupulous attributes such as intense price competition from existing or potential competitors, or the likelihood of a change in the availability of materials. Generic definitions effort to capture the effects of specific attributes in a more generalized way. Generic definitions enable designers and researchers of MCS to talk about the power of contextual variables without having to identify the scrupulous details of individual organizations. Constructing taxonomies of context and theories relating these to the use of MCS and organizational outcomes becomes more tractable. Clearly, to create prescriptive recommendations to a scrupulous organization it is necessary that the specific attributes of the environment be recognized. Moving flanked by the generic and specific should not be problematic providing the generic definitions are robust. Chapman gives a discussion of the deal-offs flanked by simplicity, accuracy and generalizability in variable definition.

The External Environment

The external environment is a powerful contextual variable that is at the base of contingency- based research. Perhaps the mainly widely researched aspect of the environment is uncertainty. Early contingency research in organization design focused on the effects of uncertainty on organizational structure. Examples contain Burns and Stalker, Galbraith, Lawrence and Lorsch and Perrow.

It is significant to distinguish uncertainty from risk. Risk is concerned with situations in which probabilities can be attached to scrupulous events occurring, whereas uncertainty defines situations in which probabilities cannot be attached and even the units of the environment may not be predictable. The importance of uncertainty as a fundamental variable in MCS contingency-based research has been stressed recently through Chapman and Hartmann. Both reinterpret characteristics of MCS research through examining the impact of environmental uncertainty.

Uncertainty and risk do not give a comprehensive account of the environment. Khandwalla gives a useful taxonomy of environmental variables. These contain turbulence (risky, unpredictable, fluctuating, ambiguous), hostility (stressful, dominating, restrictive), diversity (diversity in products, inputs, customers), complexity (rapidly developing technologies). Other units of the environment that may generate pressure or give opportunities contain complexity and dynamism, controllable and uncontrollable, ambiguity or equivocality.

Findings: external environment and MCS In accounting, uncertainty has been related to the usefulness of broad scope information and timely information; performance evaluation characterized through a more subjective evaluation approach; less reliance on incentive- based pay, non-accounting approach of performance evaluation rather than either a budget constrained or profit oriented approach and participative budgeting. Functional region, particularly research and development combined with budgetary participation was shown to enhance performance. Some proof suggests combinations of traditional budgetary controls and more interpersonal and flexible controls in circumstances of environmental uncertainty. Ezzamel accounted that high environmental uncertainty was associated with an emphasis on budgets for evaluation and required explanation of variances but also high participation and interpersonal interactions flanked by superiors and subordinates.

Merchant establish that environmental uncertainty was connected to pressure to meet financial targets but there was some flexibility through way of higher manipulation of information. In a revise of four cases, Chapman proposed that accounting has a scheduling role to play in circumstances of uncertainty but there necessity be substantial interactions flanked by accountants and other managers to cope with changing circumstances as they unfold in unpredictable ways. Environmental hostility has been associated with a strong emphasis on meeting budgets.

Hostility from intense competition has been related to a reliance on formal control and sophisticated accounting, manufacture and statistical control. Though, sure specific units of competitive location, such as strength of market location and levels in product life cycles were not associated with the importance of budgets or participation. Also, environmental complexity, self-governing of function, was associated with a reduced emphasis on budgets.

From these illustrations it can be seen that a constant steam of research in excess of the past 20 years has confirmed that uncertainty has been associated with a require for more open, externally focused, non-financial styles of MCS. Though, hostile and turbulent circumstances appear, in the main, to be best served through a reliance on formal controls and an emphasis on budgets. The question may be posed, what is the appropriate MCS for organizations operating in circumstances of uncertainty, turbulence and hostility?

The organizational design literature proposes that organizations facing extreme pressure will initially tighten control as such pressure is likely to threaten short-term survival and then adopt more organic controls. Small is recognized in relation to the appropriate design of MCS to assist in managing intricate and competing forces from the external environment.

It would be useful to look at how modern, interactive information systems can give a blend of tight controls with the opportunity to source more open, informal and subjective information. Certainly there is proof that effective organizations combine tight controls with more open, informal and flexible information and communication systems. The following propositions summarize the research findings relating MCS to the external environment. Propositions concerning the external environment and MCS:

- The more uncertain the external environment the more open and externally focused the MCS.
- The more hostile and turbulent the external environment the greater the reliance on formal controls and an emphasis on traditional budgets.
- Where MCS focused on tight financial controls are used in uncertain external environments they will be used jointly with an emphasis on flexible, interpersonal interactions.

Critical Evaluation

The distinction flanked by dimensions within the external environment, such as uncertainty, hostility and complexity is significant to MCS design. More mechanistic, formal MCS tend to give partial information in uncertain circumstances and require rapid reformulation to cope with the unfolding unpredictability. Clear specification of the environmental dimensions of interest is required, as dissimilar theories are required to consider the effects of dissimilar dimensions.

There are rich research opportunities to investigate appropriate MCS design for settings that are uncertain and also hostile and intricate. Interpreting studies that have examined the power of the external environment is complicated through the use of dissimilar events of the similar environmental construct.

For instance, Gordon and Narayanan's revise of the association flanked by perceived environmental uncertainty and more broadly scoped MCS used a measure of uncertainty that captured the intensity of competition, the dynamic and unpredictable nature of the external environment, and units of change.

In learning the similar kind of MCS variables, Chenhall and Morris used a measure of uncertainty which measured lack of information on environmental factors, inability to assign probabilities on how the environment will affect success or failure, and not knowing the outcome of decisions on how the organization would lose if the decision was incorrect.

The measure used through Gordon and Narayanan is more specifically focused on the external situation than Chenhall and Morris which has a composite of external components and implications for internal decisions. Even within the Gordon and Narayanan measure units of unpredictability are combined with difficulty.

Tymond, Stout, and Shaw give a comprehensive review of MCS research investigating the role of environmental uncertainty providing recommendations that the events should involve top managers' perceptions of the external environment. The application of a single valid and reliable measure of environmental uncertainty would assist in comparing the results of studies examining uncertainty and help build a coherent body of knowledge on the effects of this variable of MCS design. The environment will continue to be a central unit of context in contingency-based research. The specific attributes of the environment are changing and should be incorporated in future studies.

The external environment that mainly organizations face comprises increased social pressure on issues such as environmental ecology and the economic and social well being of employees and civilization. The implications for management and MCS of global competition and operations are increasingly significant.

As organizations become involved in networks involving other entities such as joint ventures and supplier and customer alliances, the boundaries flanked by what is internal and external become blurred and consequently the role of MCS will likely change. Additionally, the way in which the environment exerts pressure should be explored. Granlund and Lukka note that pressure may approach from economic causes, coercion from organizations, normative pressure derived from appropriate social conduct, and the tendency to mimic apparently successful practices.

Generic Concepts of Technology

Technology has several meanings in organization behaviour. At a common stage, technology refers to how the organization's work procedures operate (the way tasks change inputs into outputs)and comprises hardware (such as machines and apparatus), materials, people, software and knowledge. Three generic kinds of technology of importance to MCS design maybe recognized from the organizational literature: complexity, task uncertainty and interdependence.

By these notions of technology, many key attributes that may power MCS design can be derived. First, organizations producing highly dedicated, non-average, differentiated products are likely to employ intricate element/batch technologies. These will tend to involve procedures that have low analyzability of procedures and several exceptions. Also, managers are likely to have imperfect knowledge of procedures and low skill to measure outputs.

A require for flexible responses to specific customers increases interdependencies crossways the value chain involving reciprocal interactions with customers, suppliers and functional elements such as marketing, manufacture, purchasing and research and development. It might be expected that these kinds of technologies would require controls to encourage flexible responses, high stages of open communication within the workforce and systems to manage the interdependencies. Traditional, mechanistic MCS based on financial controls would not appear to suit these circumstances. Second, organizations that produce average, undifferentiated products employing capital rigorous, automated procedures are likely to employ mass manufacture and procedure technologies. These will involve highly analyzable procedures and few exceptions. Knowledge of procedures and events of output will be more readily accessible. Interdependencies are moderate being sequential.

This technology requires standardized, administrative controls such as traditional, formal financial MCS. A variant of this technology is where there are non-average products but the procedures are well understood. Interdependencies with customers are likely to be reciprocal. This technology is typical of an organization producing customized products but employing reasonably automated procedures. Controls are required that are flexible and that are constant with managing interdependencies. A reliance on traditional administrative controls, including financial MCS, is unlikely to give required flexibility and more open, informal controls will be more appropriate to manage interdependencies. At the similar time traditional, formal controls may assist in controlling procedures that are well understood.

Findings: Standardized-Automated Procedures and MCS

Technologies characterized through more standardized and automated procedures are served through more traditional formal MCS with highly dev eloped procedure controls; high (low) budget use and high (low) budgetary controls. High budgetary slack gives a buffer against low predictability within the procedures and is establish less in more predictable, automated procedures with high workflow integration. Alternatively, slack will be positively related to less automated, less predictable job/batch kind technologies.

Task Uncertainty and MCS

Technologies with high (low)task analyzability are related to a high (low)reliance on average operating procedures, programmes and plans; tasks high in difficulty and variability are associated with a low reliance on accounting performance events; knowledge of task transformations is associated with behaviour control; technologies with few exceptions that are high (low)in analyzability are associated with accounting (personnel)control s.

Mia and Chenhall demonstrated that marketing departments faced more task uncertainty than manufacture departments and consequently used broad scope information to enhance performance. Brownell and Dunk showed that there was a fit flanked by circumstances of low task difficulty, participative budgeting and a high budget emphasis; while high task difficulty suited participation with or without a strong budget emphasis. Lau et al provided alike results, although they establish that high participation and high task difficulty provided a fit irrespective of budget emphasis, while high participation and high budget emphasis enhanced performance in low task difficulty situations. Brownell and Merchant establish that higher (lower) standardization of products (high knowledge of input/output dealings) combined with flexible (static) budgets and low participation to enhance performance.

Brownell and Merchant's findings that low task uncertainty combined with more flexible budgets is somewhat inconsistent with other findings linking high task uncertainty with more informal, open MCS.

Interdependence and MCS

Low stages of interdependence have been connected to budgets, operating procedures and statistical reports; with statistical reports used for scheduling and informal coordination used in highly interdependent situations. In low interdependent public sector organizations there was an emphasis on budget analysis and managers' power on budgets but infrequent interactions with superiors and small required explanation from budgets. In more intricate situations there was less emphasis on budgets and more frequent interaction flanked by subordinates and superiors. High (low)inter dependence was establish to be associated with broad (narrow) scope MCS that focuses on appropriate aggregations and integrative information.

Strategies of customization were associated with high stages of interdependence with the latter correlating with the importance for operational decisions of the information features of integration, aggregation and timeliness. Propositions concerning generic concepts of technology and MCS:

- The more technologies are characterized through standardized and automated procedures the more formal the controls including a reliance on procedure control, and traditional budgets with less budgetary slack.
- The more technologies are characterized through high stages of task uncertainty the more informal the controls including: less reliance on average operating procedures, programmes and plans, accounting performance events, behaviour controls; higher participation in budgeting; more personal controls, clan controls, and usefulness of broad scope MCS.
- The more technologies are characterized through high stages of interdependence the more informal the controls including: fewer statistical operating procedures; more statistical scheduling reports and informal coordination; less emphasis on budgets and more frequent interactions flanked by subordinates and superiors; greater usefulness of aggregated and integrated MCS.

Modern Technologies

In excess of the past 20 years MCS research has urbanized to consider the role of advanced technologies such as Presently in Time (JIT), Total Excellence Management (TQM)and Flexible Manufacturing (FM)as dimensions of context. To set up the importance of these units of technology, accounting researchers have dawn on theories from manufacturing urbanized through theorists such Hayes et al., Skinner and from economics such as notions of complementarities as modeled through Milgrom and Roberts. Young and Selto give a review of new manufacturing practices and some implications for performance events and incentive schemes, arguing require considering technology changes within their organizational context. Notwithstanding the importance of manufacturing theories, understanding the appropriate fit flanked by MCS and advanced technologies is assisted through reflecting on the vital, generic notions of technology addressed above.

Kalagnanam and Lindsay argue that JIT is best suited to open, informal, organic shapes of controls. They claim that organic systems can best manage the secure linkages or coupling within JIT that can cause variability (task uncertainty due to several exceptions) between units of manufacture procedures (interdependence). Organic systems are also required to manage require for flexible responses to customers, which involves coordinating reciprocal interdependencies crossways the value chain.

Finally, JIT implies continuous improvement that is best served through commitment to change from the shop floor which is encouraged through organic systems. Alike arguments may be made for implementing innovative MCS in TQM and FM situations.

These technologies have high variability and low analyzability. The low analyzability derives from require to continually use potential complementarities flanked by the several units of TQM practices.

Also, TQM and FM involve the effective management of interdependencies within manufacture procedures including relationships with customers, suppliers and other external parties.

Controls are required to encourage managers and shop floor employees to focus on the critical units of variability within the TQM programmes and to give effective links crossways the value chain.

This information is generated at both the procedure (cybernetic kind controls such as statistical procedure controls)and strategic stages (*i.e.* linking procedures to strategic outcomes). Continuous improvement requires access to knowledge on world's best practice and systems to encourage innovation. Appropriate control systems should be open and informal, contain broad scope information, benchmarking, and performance events that indicate links flanked by strategy and operations such as balanced scorecards and strategic integrative controls.

Findings: Advanced Technologies and MCS

Ittner and Larcker demonst rated that product focused TQM was connected to timely problem solving information and flexible revisions to reward systems. They establish for advanced (holistic) TQM, external benchmarking and the integration of excellence and strategic information is significant. Ittner and Larcker examined the association flanked by excellence programmes and a diversity of strategic controls related to implementation, internal and external monitoring.

Links flanked by excellence and strategic controls were established, with differences flanked by countries. Also, example-wide performance effects were restricted to controls concerning manager's participation in approving excellence programmes and team formulation, with other associations contingent on industry effects. Sim and Killough establish that customer and excellence performance was higher in TQM and JIT situations where there was customer and excellence related performance goals and incentives compared to where fixed pay was used. Ittner, Larcker, Nagar, and Rajan reported that performance gains from supplier partnership practices were associated with extensive use of non-price selection criteria, frequent meetings and interactions with suppliers and supplier certification.

These controls were not effective for arms length supplier dealings. Kalagnanam and Lindsay establish that organic MCS were associated with effective JIT systems. Some studies have examined the role of non-financial performance events in advanced technologies. Banker et al. establish that JIT, excellence and teamwork were associated with the provision of non-financial, excellence and productivity events to shop floor employees. There is some proof suggesting that relying on non-financial events to evaluate managers in TQM situations gives interactive strategic control. Mia establishes that the provision of broadly based MCS enhanced organizational performance in JIT settings.

The broad MCS incorporated performance targets related to non-financial manufacturing indicators, actual performance on those targets, organizational financial indicators and industry and organizational trends on overall performance. Customer focused manufacturing, jointly with advanced manufacturing technology (AMT), have been associated with non-financial events. It is noteworthy that there is ambiguity in findings related to the extent to which associations flanked by usefulness of non-financial performance events and advanced technologies are related to enhanced performance. For instance, Chenhall establish positive performance effects flanked by combinations of non-financial events and TQM, while Perera et al. did not. One explanation for these differing findings is in the use of the performance events.

Chenhall related the events to reward and compensation systems, whereas Perera et al. did not create this linkage. Perhaps the extent to which non-financial events are used to evaluate and reward managers may be significant in understanding links flanked by performance events, advanced technologies and performance. This suggestion is constant with Sim and Killough's findings that incentive pay enhanced the positive effects of TQM and JIT on customer and excellence performance. Also, Larcker's market -based revise establish that the combination of incentive schemes and capital investment was associated with improved investor return.

A recent revise, reporting a laboratory revise, demonstrated the importance of incentive schemes to enhance both absolute performance and rates of improvement through encouraging individuals to spend more time on tasks and to use and examine information. Foster and Horngren found that flexible manufacturing systems (FMS) were associated with performance events focused on time, excellence, operating efficiency and flexibility.

There was also a change in the costing methods (allocations, treatment of costs as era and changes in the components of direct costs). Though, flexible manufacturing (FM)has been connected to a de-emphasis of efficiency-based events with control derived from integrative liaison devices. It is to be noted that there is a variation flanked by FMS which are technological systems such as computer assisted design and computer assisted manufacturing (CAD/CAM) and FM which is a generic notion of technology emphasizing a strategy of flexible response and customization. More research is needed to explore whether both focused formal controls at the operational stage and more intricate integrative devices can coexist to assist control within TQM and FM situations.

Also, links flanked by dissimilar kinds of controls for operational, managerial and strategic decisions should be explored. For instance, Chenhall and Langfield-Smith (1998b)link performance with combinations of several traditional and modern controls with a range of strategies and manufacturing practices. Propositions concerning advanced technologies and MCS:

- TQM is associated with broadly based MCS including timely, flexible, externally focused information; secure interactions flanked by advanced technologies and strategy; and non-financial performance measurement.
- The extent to which combinations of advanced technologies and non-financial performance events are associated with enhanced performance depends on the degree to which the events are used as part of reward and compensation schemes.
- The advanced technologies of JIT and FMS are associated with broadly based MCS such as informal controls and greater use of non-financial performance events.
- FM is associated with the use of informal, integrative mechanisms.
- Supplier partnership practices are associated with non-financial events, informal meetings and interactions crossways the value chain.

Critical Evaluation

The three generic concepts of technology that have been used in MCS research (complexity, task uncertainty and interdependence)a re separate constructs but there are some general themes concerning uncertainty. It appears likely that conversion of inputs into outputs within less intricate, mass manufacture technologies is more programmable and predictable than in job or batch styled technologies servicing customized products. High stages of predictability are associated with the throughput of procedure technologies but not for the management of breakdowns and other exceptions.

The construct of task uncertainty concerns lack of information and is a combination of variability or lack of knowledge in relation to the alternatives and uncertainty in relation to the how to examine the variations, or measure outputs, that happen throughout the conversion of materials into output. A higher stage of interdependence, where the work of one sub-element is complicated through having to rely on another, raises the possibility of more uncertainty derived from lack of control in excess of the supplying sub-element.

The importance of uncertainty as an aspect of both environment and technology has lead to some ambiguity flanked by environmental and technological uncertainty in MCS research. For instance, Hirst argued that accounting performance events would be inappropriate in circumstances of environmental uncertainty but measured uncertainty with a composite measure comprising both units of task and environmental uncertainty arguing that the concepts are measuring the similar thing. Ross theorizes effects flanked by task uncertainty and performance events but uses events of environmental uncertainty.

Clarification of links flanked by environmental and technological uncertainty is required to isolate potentially dissimilar effects of these variables on MCS design.

For instance, external uncertainty implies a lack of information which creates it hard to plan kinds of products and services, stages of output and make contingency plans. Also, it creates evaluation hard as demand may change in ways beyond the control of managers. This suggests that more flexible, interactive MCS are required to encourage learning and version and evaluate managers on the foundation of more subjective events or against adjustable criteria dependent on changing circumstances.

The uncertainty associated with technology is, in part, derived from the environment with the technology being responsive to the uncertainty associated with markets and product necessities. Therefore, technology may respond to environmental uncertainty through becoming more flexible or through employing JIT techniques. The appropriate MCS design is likely to be more flexible and organic. Though, uncertainty, also, is caused directly within the technological procedures, independently from environmental circumstances.

This may be derived from a search for improvements in product design and cost efficiencies and is likely to augment concern with managing uncertainty and complexity associated with the manufacture procedures. These circumstances may prompt the adoption of scheduling and evaluation systems such as action- based accounting, non-financial manufacturing performance events and supplier networks. Despite the links flanked by environmental and task uncertainty, where possible researchers should draw on work that has tried to resolve issues related to the validity and reliability of events concerning these contextual variables. An instance of this is Brownell and Dunk who sought to reconcile findings related to the role of task uncertainty in the revise of budgetary related behaviours.

Studies through Brownell and Hirst and Hirst used a measure of task uncertainty that aggregated the separate dimensions of task difficulty (analyzability)and variability (number of exceptions). Brownell and Dunk argued that such a composite measure is inappropriate as it mixes up the potential effects of difficulty and analyzability. They establish that task difficulty and not task variability moderated the effects of budget behaviours on performance.

The region of modern manufacturing practices, such as JIT, TQM, FM and AMT, has provided several opportunities for contingency-based research. Ideas from economics concerning complementarities are likely to prove useful in modeling the way multiple characteristics of manufacturing can be combined optimally.

Developing an understanding of best manufacturing practices and the way in which manufacturing aligns with or gives the impetus for strategy would appear to be a necessary step in ensuring that MCS design maintains relevance to the technological core of organizations. Closer cooperation flanked by MCS researchers and manufacturing technology experts and industrial engineers would be fruitful. The importance of advances in IT cannot be underestimated. The adoption of interactive IT systems, such as SAP R/3, often triggers the adoption of scrupulous performance and costing systems. As the average life span of products decrease, consideration of the life cycle of products has become a concern in manufacturing. Short product life cycles lay demands for new product initiatives and alter cost structures.

Also, decreasing life cycles augment operating risk and require increased capital investment. Understanding how MCS innovations, such as target costing, can assist management within these settings will likely become increasingly significant. There has been small work that has investigated how MCS are best suited to dissimilar levels in the growth of firms. Significant topics are the role of more formal systems at the levels of new firm formation, early growth, maturity and decline. Questions arise concerning the necessities of MCS at these dissimilar levels, particularly the extent to which MCS can assist in the transition from early growth to more mature levels.

Moores and Yuen provide an examination of issues concerning dissimilar characteristics of MCS that are significant for dissimilar levels of the growth cycle of firms. Finally, it is noteworthy, that mainly contingency based MCS research has involved big, manufacturing organizations. There have been some studies in the hospital and hospitality sectors but, on the whole, there has been small research investigating the service and government sectors. Some examples contain studies within government agencies, hospitals, research and development and marketing departments. The growth in importance of service industries such as hospitality and tourism, and the introduction of managerial approaches to public sector management give several opportunities for future research.

Organizational Structure

Organizational structure is in relation to the formal specification of dissimilar roles for organizational members, or tasks for clusters, to ensure that the behaviours of the organization are accepted out. Structural arrangements power the efficiency of work, the motivation of individuals, information flows and control systems and can help form the future of the organization.

There have been several definitions of organizational structure. A significant distinction is the variation flanked by the outcomes of structure and the structural mechanisms. Lawrence and Lorsch refer to structure, generically, as the way in which the organization is differentiated and integrated. Differentiation is concerned with the extent to which sub-element manager's act as quasi entrepreneurs, while integration is defined as the extent to which the subunits act in ways that are constant with organizational goals. The mechanisms to achieve differentiation involve decentralizing power, while integration involves rules, operating procedures, committees and the like.

Pugh, Hickson, and Turner and Pugh, Hickson, Hinings, and Turner empirically recognized examples of structural mechanisms that have been used commonly in contingency-based research, including centralization, standardization, formalization and configuration. Burns and Stalker discus s structure, generically, in conditions of mechanistic and organic approaches. The means to achieve these shapes of structure involve mechanisms such as rules, procedures and openness of communications and decision procedures.

Perrow identified structure in conditions of bureaucratic and non-bureaucratic approaches. Designers of MCS have been concerned with formulating MCS to be constant with the intent of organizational structure. Consequently, it is useful to consider the extent to which MCS are mechanistic or organic, or to which they differentiate or integrate. The choice of structure in organizational contingency research has focused on the appropriate structure to fit the stages of uncertainty in the environment, strategy and the organization's technology.

Usually, it is whispered that more organic structures are suited to uncertain environments. Though, it should be noted that Lawrence and Lorsch recognized require for higher stages of differentiation to cope with diverse and uncertain environments and that this causes potential integration troubles which require sophisticated liaison mechanisms (integrative personnel, meetings), rather than rules and procedures. This kind of response is something of a hybrid flanked by mechanistic (for differentiation)and organic kinds of structure (for integration)to manage uncertainty. A big body of literature suggests that strategies characterized through diversification require differentiated, divisional structures.

Also, it may be argued that once scrupulous structures are in lay then decisions will be influenced through the opportunities afforded managers from power granted to them and, perhaps, the political interests of those individuals. Therefore, strategy might follow structure. Often the structural arrangements have significant implications for information flows that may form or bias the future directions of the organization.

In the prior part extensive links flanked by technology and kinds of controls were drawn. It is, also, noteworthy that early studies of organizational design recognized significant links flanked by technology and structure. Particularly, early research establishes that changing to more efficient technologies did not necessarily lead to enhanced effectiveness accepted. Implementing the technically efficient technologies required reformulating roles and structures with dynamic negative effects on the way individuals related to the new technologies, and consequently, deterioration in performance. It was apparent that socio-technological approaches were required to ensure improved organizational performance. These early observations are significant to several recent structural innovations such as work-based team that effort to harness growths in technology with the efficient blending of appropriate skills and the motivating force of teamwork. When evaluating contingency relationships flanked by MCS and structure, units of environment, technology and strategy are likely to be implicated in the relationships and, as such, much can be gained through considering them at the similar time.

Findings: Organizational Structure and MCS

Big firms with sophisticated technologies that are decentralized have been associated with a strong emphasis on formal MCS; and big, diverse, decentralized firms used more administrative controls. Decentralized organizations were recognized as perceiving aggregated and integrated information as useful. Managerial performance was associated with the interaction flanked by decentralization and each of the features of broad scope, integrated, aggregated and timely information received/obtained from MCS.

There is some proof linking MAS to functional differentiation. Functional differentiation (more responsibility in excess of regions of manufacturing)was connected to formality of budgetary procedures. Hayes establishes that the importance of evaluating factors related to internal operations, external circumstances and interdependencies depended on the functional nature of departments.

In manufacture departments, overall effectiveness was associated with factors related to the performance of internal factors. For marketing, performance of factors related to the external operating circumstances and interdependencies were mainly significant. Functional differentiation has been connected to environmental uncertainty to demonstrate how research and development elements, compared to marketing, are better suited to participative budgeting.

Mia and Chenhall establish that marketing, compared to manufacture, involves higher task uncertainty that explained why marketing managers used broad scope information more effectively than those in manufacture. Concerning scrupulous functional decisions, Foster and Gupta recognized that improvements in MCS would be valued for pricing decisions, customer mix, sales force/promotions and product mix. Costing information was perceived as useful for decisions concerning products and customers. There was a variation flanked by potential and actual use of MCS in the region of marketing. Budgetary participation has been studied extensively and associated with a wide diversity of contextual units. Structural contingencies connected to effective participative budgeting have incorporated functional differentiation, specifically research and development compared to marketing; leadership approach employing high compared to low budget emphasis; a consideration rather than initiating approach of leadership; as well as the findings related to decentralization.

As noted, the theories used to look at functional differentiation relied on links to external environmental uncertainty, rather than structure, per se. The ways in which MCS combine with units of organizational structure to give differentiation and integration within modern organizational structures give several opportunities for worthwhile research. Particularly, there are few studies that have measured the fit flanked by organic structures and MCS. Organizational theory would suggest require for flexible, open information systems rather than tight budgetary systems.

Gordon and Narayanan establish that organic structures were best served through broad scope and future oriented information. Some researchers have established that more organic, behaviourally oriented implementation is required to ensure the success of action-based accounting. Gosselin establish that action-based costing is adopted and implemented in organizations with more mechanistic structures.

Particularly, mechanistic structures (vertical differentiation or bureaucratic decision procedures)facilitate adoption of activity based costing (an administrative innovation)and centralization and formalization were associated with implementing action-based costing. Organic structures were more suited to action analysis and action-cost analysis (technological innovation). Presumably, organizations proceeding from action analysis to action-based costing would require units of organic and mechanistic structures to carry them through the levels of action analysis to action-based costing.

A significant unit of modern structures is teams. As yet there are few studies that have measured the role of MCS within team based structures. Young and Selto predicted that teamwork and problem solving abilities of shop floor employees would be associated with high performance related to JIT outcomes.

Their revise in a single organization did not reveal these associations due, in part, to an inability of workers to address procedure troubles and poor implementation of JIT-compatible management controls. Scott and Tiessen statement that team based structures were associated with high task complexity and that team performance was associated with use of comprehensive performance events (financial and non-financial), formulated participatively and used for compensation. In an experimental revise, Drake, Haka, and Ravenscroft establish that in team structures the interaction flanked by ABC and rewards based on group incentives was associated with cooperative innovations, lower costs and higher profits. Propositions concerning organizational structure and MCS:

- Big organizations with sophisticated technologies and high diversity that have more decentralized structures are associated with more formal, traditional MCS (*e.g.* budgets, formal communications).
- Research and development departments compared to marketing departments, which face higher stages of task uncertainty, are associated with participative budgeting; and marketing compared to manufacture departments, which face higher stages of external environmental uncertainty, are associated with more open, informal MCS.
- The structural features of functional differentiation based on research and development compared to marketing, leadership approach characterized through a consideration compared to initiating approach, and higher stages of decentralization are associated with participative budgeting.
- Decentralization is associated with the MCS features of aggregation and integration.
- Team based structures are associated with participation and comprehensive performance events used for compensation.
- Organic organizational structures are associated with perceptions that future orientated MCS are more useful, and with the effective implementation of action analysis and action-cost analysis.

Critical Evaluation

Structural mechanisms have been conceived of as involving differentiation and integration. Concerning differentiation, conventional thinking in management accounting proposes that decentralization should be combined with profit center responsibility accounting systems. To achieve integration easy mechanisms, such as operating procedures and formal budgets, have been recommended.

It is of interest to observe the extent to which these recommendations appear somewhat inconsistent with the suggestions of Lawrence and Lorsch that highly differentiated organizations should employ intricate liaison mechanisms to achieve integration. Closer inspection of empirical findings suggests that comprehensive and formal mechanistic controls might be only one aspect of coordinative efforts in differentiated organizations. Khandwalla establish that big decentralized companies employed sophisticated controls but also utilized high stages of participation and human dealings approaches to coordinate behaviours. Certainly, participation in budgeting has been connected to decentralized organizations. Merchant establish participation was one aspect of administrative controls. Gul, Tsui, Fong, and Kwok establish an association flanked by decentralization and participative budgeting.

How the participation of individuals in formal budgets might link to more organic shapes of control is a motivating region for further research. The role of budgets within organizations that have urbanized structures based on delayering, developing teams and empowering employees should be investigated. Galbraith allude s to the require to focus on the procedure of decision creation and disagreement resolution in situations in which there is ambiguity and disagreement flanked by the several structural elements and roles within organizations..

Care should be employed in selecting measurement instruments related to structure. Structure has been measured in conditions of decentralization of power, structuring of behaviours, interdependence and organic-mechanistic orientations. Events of decentralization, structuring of behaviours and interdependence have relied, in the main, on those urbanized through the Aston school. The organic-mechanistic nature of structure has been derived from Khandwalla.

The Aston events have been subjected to considerable scrutiny and empirical testing for validity and reliability in the organizational literature. The use of more novel events, such as those related to team-based structures, will require consideration of work that has urbanized these events. As with other units of context, in modern settings, structure remnants a significant factor in understanding MCS design. Several argue that adjustments to structure are required to ensure employee commitment to organizational goals related to continuous improvement.

Structural innovations, such as delayering, flat structures, networking, procedure orientations and team-based work clusters concern the removal of barriers flanked by organizational behaviours. Such seamless organizational structures appear to be inconsistent with traditional profit centers and responsibility accounting, yet several organizations uphold these hierarchical structures. Empowerment and teamwork have replaced participation as the appropriate concept for understanding the efforts of several organizations to gain employee involvement. Issues of coordination, performance evaluation and reward systems in team-based organizations are significant research regions. Much can be learned from linking MCS research agendas with work of human resource management researchers.

Size

Growth in size has enabled firms to improve efficiency, providing opportunities for specialization and the division of labour. Big organizations tend to have more power in controlling their operating environment, and when employing large-scale mass manufacture techniques have reduced task uncertainty. Though, as an organization becomes superior the require for managers to handle greater quantities of information increases to a point where they have to institute controls such as rules, documentation, specialization of roles and functions, extended hierarchies and greater decentralization down hierarchical structures. Modern big organizations often develop secure associations with suppliers and customers, which blur the boundaries flanked by organizations, thereby rising further the size of the entity.

Size has also provided organizations with the possessions to expand into global operations, sometimes through way of mergers, takeovers, licensing or other collaborative arrangements. These growths make additional administrative concerns due to increased stages of complexity within the manufacture procedures and with managing interdependencies with global partners.

Findings: Size and MCS

Few MCS studies have explicitly measured size as a contextual variable. In the main, studies have examined relatively big organizations, usually justifying this as it is big firms that tend to adopt the kind of practices incorporated within more formal MCS. Studies that have examined size have measured its effects jointly with other units of context such as technology, product diversity and have examined an array of controls.

Khandwalla (1972, 1977)establish that big firms were more diversified in product rows, employed mass manufacture techniques, were more divisionalized and made greater use of sophisticated controls and environmental information gathering such as forecasting and market research. The papers through Bruns and Waterhouse and Merchant, discussed earlier in conditions of organizational structure, give proof related to size.

Bruns and Waterhouse identified two shapes of control associated with size: administrative with big firms and personal with little firms. Administrative control comprised more sophisticated technologies, formalized operating procedures, high stages of specialists and work related rules. Managers perceived that employees had high stages of control and had high stages of participation in setting standards and spent more time in budgeting.

They perceived budgets as limiting innovation and flexibility in structuring organizations. Interpersonal control involved centralized decision creation, individuals saw themselves as having more interaction on budget related matters, not having their methods of reaching budgets accepted and being required to explain budget variances. Individuals were satisfied with their superior–subordinate relationships. Merchant's revise also measured size as an aspect of a multiple variable approach. Big, diverse firms were more decentralized, used sophisticated budgets in a participative way and employed more formal communications. Propositions concerning size and MCS

- Big organizations are associated with more diversified operations, formalization of procedures and specialization of functions.
- Big organizations are associated with more divisionalized organizational structures.
- Big size is associated with an emphasis on and participation in budgets and sophisticated controls.

Critical Evaluation

Mainly contingency-based MCS research has studied superior organizations but has not measured size difference within superior entities. This is unfortunate as there is proof from early organizational contingency studies that the connection flanked by size and administrative arrangements such as specialization, formalization and the vertical span increases with size but at a declining rate. Therefore, while it is reasonable to assume that big firms employ formal MCS, it is possible that dissimilar kinds of controls will be appropriate within these big firms, depending on size. The role of MCS in smaller or medium sized entities has not received small attention in the contingency- based MCS literature, nor has the role of MCS in firms that change size due to rapid internal growth, take-in excess of or merger been explored.

It appears likely that the role of formal and interpersonal controls would differ depending on size and rate of change in size. Several opportunities for MCS, contingency-based research are likely to be establishing in the region of little and medium sized business. An impact of technological change and structural reform has been to reduce the number of employees, both shop-floor employees and the number of middle stage managers. In as much as the number of employees is associated with coordination and control issues, reduced size, due to the substitution of capital for labour, will have implications for MCS.

For instance, the combination of procedure controls to monitor machines and informal controls for evaluating people will likely become more significant where there are fewer employees operating and managing capital rigorous technologies. Concerning measurement, there are many ways of estimating size including profits, sales volume, assets, share valuation and employees. The use of financial events can create comparisons flanked by organizations hard as dissimilar accounting treatments can be establish flanked by firms. Mainly contingency-based MCS studies have defined and measured size as the number of employees. Numbers of employees has been establishing to correlate with net assets. It is possible that the precise measure of size could be significant depending on the unit of context and dimensions of MCS being studied.

If the theory is considering the effectiveness of budgets to coordinate individuals behaviours then employees is appropriate. If, though, the revise is examining the effects of environment on the effectiveness of customer focused accounting then sales and assets might be more appropriate as these events capture market power that can lead to barriers to entry or industry concentration. Khandwalla argues that forecast sales are the best indicator or how size may relate to scheduling, budgeting and structural modification.

Strategy

Strategy is somewhat dissimilar from other contingency variables. In a sense it is not a unit of context, rather it is the means whereby managers can power the nature of the external environment, the technologies of the organization, the structural arrangements and the control civilization and the MCS. The role of strategy is significant as it addresses the criticism that contingency based research assumes that an organization's MCS is determined through context and that managers are captured through their operating situation.

Recently, MCS research has recognized that managers have 'strategic choice' whereby they can location their organizations in scrupulous environments. Therefore, if the current product range is too uncertain, reformulating product strategy into a market that is more predictable may remove the pressure from the environment. It may, also, limit potential opportunities and so require the organization to look at its attitudes to the tradeoff flanked by potential returns and acceptable risk and uncertainty. Notwithstanding the strategic direction selected through the organization, contingency- based research predicts that sure kinds of MCS will be more suited to scrupulous strategies.

The powerful power of strategy is evidenced through the popular use of conditions such as strategies of TQM, the strategic imperative of an empowered workforce and strategic management accounting. Langfield-Smith gives a summary of research into MCS and strategy. Much generic taxonomy has been urbanized including entrepreneurial-conservative; prospecters-analysers-defenders; build-hold-harvest; and product differentiation- cost leadership.

Proof from the strategy-organizational design research suggests that strategies characterized through a conservative orientation, defenders, harvest and cost leadership are best served through centralized control systems, dedicated and formalized work, easy co-ordination mechanisms and attention directing to problem regions. Strategies characterized through an entrepreneurial orientation, prospectors, build and product differentiation are connected to lack of standardized procedures, decentralized and results oriented evaluation, flexible structures and procedures, intricate co-ordination of overlapping project teams, and attention directing to curb excess innovation. Simons argues that four dimensions of MCS link to strategy: belief systems to communicate and reinforce vital values and missions, boundary systems to set up limits and rules to be respected, diagnostic controls to monitor outcomes and correct deviations, and interactive controls to enable top managers to personally involve themselves with subordinates and operations with a view to forcing dialogue and learning.

Findings: Strategy and MCS

From MCS research, proof suggests links flanked by strategy and cost control and to formality of performance evaluation. The studies are focused on strategy at the strategic business element stage, rather than corporate or functional stages. Mainly of the studies explore the association flanked by MCS and strategic typologies. Conservatives, defenders and cost leadership strategies discover cost control and specific operating goals and budgets more appropriate than entrepreneurs, prospectors and product differentiation strategies. Simons establish that entities with small sense of urgency in relation to the creating a vision, did not employ interactive controls.

These generalizations are fairly simplistic. Merchant establish no association flanked by dissimilar growth strategies and pressure to meet financial targets. Simons demonstrated that tight controls were apparent in more entrepreneurial strategies, perhaps to balance excessive innovation and to help learning in uncertain environments. Chenhall and Morris found that tight control was appropriate for conservative strategies; though, tight control was also establish in entrepreneurial situations but, importantly, operating jointly with organic decision styles and communications. Again, the apparent paradox can be explained through require for organic systems to encourage innovation and tight control to curb excessive innovation. Concerning performance measurement, build compared to harvest strategies, which involve low specialization and difficulty in measuring outcomes, suit more subjective and extensive-condition controls *e.g.* Product differentiation was associated with a low emphasis on meeting budgets, and when combined with high service sharing was associated with a reliance on more subjective behavioural control.

A revise through Abernethy and Brownell establish that hospitals undergoing strategic change (a more prospector kind of strategy) used budgets interactively, focusing on dialogue, communication and learning (more organic styles of control). Van der Stede showed that product differentiation strategies were associated with less rigid budgetary control, which in turn, was associated with increased budgetary slack, although there were no direct effects flanked by strategy and slack. Proof on the usefulness of more broad scope scheduling information for prospector companies and for those following build compared to harvest strategies was establish through Guilding.

In this revise, the scope of the information related to competitor-focused accounting which incorporated competitor cost assessment, competitive location monitoring, competitor appraisal based on published financial statements, strategic costing and strategic pricing.

Bouwens and Abernethy found that the stage of importance to operational decision creation of more integrated, aggregated and timely information was correlated with customization strategies.

While associations with broad scope information were not establish, the revise focused on importance for 'operational' decisions, which presumably excluded decisions concerning markets and customer necessities which are more likely to involve broad scope information. Propositions concerning strategy and MCS:

- Strategies characterized through conservatism, defender orientations and cost leadership are more associated with formal, traditional MCS focused on cost control, specific operating goals and budgets and rigid budget controls, than entrepreneurial, build and product differentiation strategies.
- Concerning product differentiation, competitor focused strategies are associated with broad scope MCS for scheduling purposes, and customization strategies are associated with aggregated, integrated and timely MCS for operational decisions.
- Entrepreneurial strategies with both formal, traditional MCS and organic decision creation and communications.
- Strategies characterized through defender and harvest orientations and following cost leadership are associated with formal performance measurement systems including objective budget performance targets, compared to more prospector strategies which require informal, open MCS characterized through more subjective extensive condition controls and interactive use of budgets focused on informal communications.

Critical Evaluation

Ideally, the role of strategy is dynamic involving managers in continually assessing the way combinations of environmental circumstances, technologies and structures enhance performance. MCS has the potential to aid managers in this procedure through assisting them in formulating strategy related to markets and products, required technologies and appropriate structures. MCS can then be implicated in the implementation and monitoring of strategies, providing feedback for learning and information to be used interactively to formulate strategy.

Few studies in MCS have investigated these issues, rather mainly have been restricted to identifying MCS that are appropriate for dissimilar strategic archetypes. While there are some general units in these dissimilar strategic archetypes, there are important differences, consequently care is needed in developing theory that is specific to the archetypes employed in the revise. For instance, Fisher and Govindarajan develop theory to look at strategy and alternative controls based on the dissimilar requires derived from combinations of strategic mission, by concepts of build, hold and harvest, and competitive strategy, by product differentiation and low cost taxonomies. The extent to which these archetypes, which were urbanized in the 1970s and 1980s, uphold their relevance to modern settings is questionable.

Strategies are being complicated through require for mainly organizations to be both low cost producers and to give customers with high excellence, timely and reliable delivery. More meaningful associations flanked by strategy, environment and internal operations may become apparent if specific units of strategic priorities are investigated. Relevant research is accessible based on modern strategic priorities and has been applied in management accounting research. The role of MCS in organizations involved in strategic change is a significant topic. Some insights can be gained through examining movement crossways dimensions of archetypes, such as a change from harvest to prospector orientations, Though, greater understanding of how MCS is implicated in strategic change is possible through considering theories that relate to differences flanked by incremental, synthetic and discontinuous change, the role of strategic intent and strategic possessions, the variation flanked by planned and emerging strategies, styles of management that encourage change, the impediments to change of any formal resource allocation procedure, and the way MCS can be used to manage both evolutionary and revolutionary change.

There have been concerns with the measurement of strategy. Events used to revise strategy have been criticized as mixing up units of the environment with organizational attributes (therefore studies of strategy and environment would be invalid). Events tend not to relate to competitors which create comparisons crossways industry clusters problematic. Managers have difficulty relating to descriptions used to capture generic typologies such as build, harvest and prospect. Strategy research should consider work that has attempted to validate strategy events such as Dess and Davis, Kotha et al., Miller and Friesen, Miller and Roth and Shortell and Zajac.

Civilization

The connection flanked by the design of MCS and national civilization symbolizes an extension of contingency-based research from its organizational foundations into more sociological concerns. The vital proposition is that dissimilar countries possess scrupulous cultural features. This predisposes individuals from within these cultures to respond in distinctive ways to MCS. Civilization has becomes significant in the design of MCS, in excess of the past 20 years, as several companies have urbanized multi-national operations. These companies face the issue of whether to transfer their domestic MCS overseas, or redesign their systems to fit the cultural features of the offshore entities.

Compared to studies of other contextual variables, research into civilization has been limited and is somewhat exploratory. There is a plethora of meanings of civilization. Though, Kaplan claims there is consensus in the middle of anthropologists that civilization is composed of patterned and interrelated traditions, which are transmitted in excess of time and legroom through non-biological mechanisms based on man's uniquely urbanized linguistic and non-linguistic symbolizing capabilities. Civilization can be described through inherent traits such as knowledge, belief, art, morals, law, custom, and other capabilities and habits acquired through man as a member of civilization.

Though, often civilization is conceptualized as a set of features in accessible to suit the methodological and scientific requires of the research society. The mainly widely used features were urbanized through Hofstede who described the cultural values as power aloofness (acceptance of unequal sharing of power), individualism vs. collectivism (placing self-interest ahead of the group), uncertainty avoidance (preference to avoid uncertainty and rely on rules and structures), masculinity vs. femininity (attainment, assertiveness and material success vs. modesty and preference for excellence of life)an d, subsequently, confucian dynamism. Virtually all MCS contingency-based studies have used these values to revise the power of civilization.

Findings: Civilization and MCS

Contingency-based research in MCS has examined associations flanked by cultural dimensions and units of structure such as standardization, decentralization, and control system features such as formality on controls, reliance of accounting performance events and budgetary participation. Overall, the research has provided mixed results as to whether civilization does have effects crossways characteristics of MCS. There are few regions where consensus can be drawn. This is because studies have examined dissimilar combinations of cultural dimensions and have measured characteristics of MCS in dissimilar ways.

As a consequence there is small overlap flanked by studies to enable themes to be drawn or comparisons made and generalizations urbanized. The following are examples of studies that have examined accounting controls. Harrison demonstrated that differences flanked by Singapore and Australian managers did not moderate the connection flanked by budget emphasis in evaluation and either job related tension or job satisfaction. Though, the connection flanked by reliance on accounting performance events and low job related tension and high job satisfaction was stronger for Singapore managers, the explanation being that these managers had low individualism and high power aloofness compared to Australian managers. O'Conner argued that the low power aloofness establish in western parent companies would control in excess of high power aloofness establish in their local Singapore subsidiaries, thereby enhancing the effectiveness of the parent MCS.

By these arguments he establishes that the connection flanked by role ambiguity and superior/subordinate relationships (perceptions of competence and trustworthiness) and both participation in budgeting and in performance evaluation was stronger in foreign subsidiaries than local Singapore entities. Merchant, Chow, and Wu studied Taiwanese and US firms and establish that civilization was not significant in explaining use or effectiveness of the degree of subjectivity in profit center manager's performance evaluation. Only use of extensive-condition incentives was more significant in Taiwanese firms.

Many studies have measured broader characteristics of MCS with less equivocal results. Snodgrass and Grant establish that Japanese, compared to US, companies experience less explicit controls and more implicit controls in monitoring, evaluation and rewarding. Ueno and Wu also establish differences flanked by Japanese and US managers on MCS features. They theoretically connected individualism with US managers and establish, empirically, that they used more formal communications, built slack, used controllability in budgeting and extensive-condition horizons for performance evaluation.

Uncertainty avoidance was connected to Japanese firms to explain a preference for broad time horizons and structured budgetary procedures. These associations were not supported empirically leading to the conclusion that individualism is the dominant predictor of MCS. Vance, McClaine, Boje and Level studied formality of controls, team development, appraisal systems, intrinsic/extrinsic rewards and frequency of feedback in Indonesian, Malaysian, Thai and US firms. Important differences were establish flanked by US and Asian firms but also flanked by the dissimilar Asian firms. This revise is distinctive as it used both Hofstede's dimensions plus other concepts of civilization drawn from anthropology. Finally, studies by experimental methods have failed to support expected effects and have revealed ambiguous findings. Given the exploratory nature of research examining civilization and the lack of consensus on findings, only a common proposition relating civilization to MCS is presented. Proposition concerning civilization and MCS:

- National civilization is associated with the design of MCS.

Critical Evaluation

The dominant notion of civilization employed in MCS-civilization research has been the Hofestede values. Though, many criticisms can be made as to how this approach to defining and measuring civilization has been employed. First, it assumes that the dissimilar values have the similar intensity within a civilization.

If this is not the case, then some value may be more dominant than others and have a prominent effect. Second, some studies do not consider all of Hofstede's values. It is possible that omitted values may have effects that are relevant to the revise.

Third, mainly studies assume that countries differ on values and proceed to test for differences flanked by countries without directly assessing cultural values. It is, so, significant to check that the assumed values of a country are still apparent in modern studies. Fourth, while Hofstede's values give a convenient tool for research, it does symbolize a restricted view of civilization. Its exclusive use has prohibited development of understanding how behaviour is influenced through the fundamental traits that power how individuals think, feel and respond. More subtle notions of civilization involving myths and ritual, language and narrative are not measured. It appears likely that theories and methods drawn from anthropology and sociology are more suited to understanding how these subtle factors combine to power how individuals respond to MCS.

While national civilization has been studied extensively, it appears likely that other variables such as markets and technologies may interact with cultures in systematic ways to affect MCS design. For instance, the adoption of sure kinds of advanced technologies appears to work mainly effectively if attributes of collectivism are apparent.

This combination of technology and civilization may suggest that sure kinds of performance events, suited to the technology but constant with collectivism, would be more appropriate.

One variable that offers promise in the revise of civilization is organizational civilization. It is possible that a strong organizational civilization may control national civilization in the work situation. Small work has been completed in the region of organizational civilization and MCS design. As with national civilization, the meaning of organizational civilization and its revise are well served through the application of the research paradigms and methods from sociology and anthropology.

Continuing Relevance of Traditional Units of Context

Insights into the present-day context of MCS can be gained through reflecting on the issues drawn from traditional contingency-based work. The environment will become more uncertain, hostile and intricate as a result of modern pressures. There will be requiring for organizations to develop increased environmental responsibility. Technologies will be established to have varying degrees of complexity, uncertainty and interdependencies that promote control issues. Structures will be employed that assist in developing more organic ways to communicate, but also give enhanced differentiation to motivate and location individuals secure to the business operations.

Additionally, structures will be sought that empower individuals, with the purpose of providing a healthy and fulfilling work environment while better equipping the organization to achieve best practices. The challenges to coordination derived from size will increasingly become significant as organizations enlarge due to developing secure relationships with suppliers and customers and engaging in global operations through direct expansion, acquisition and merger. Notions of strategy are likely to be redefined and it will be necessary for MCS researchers to stay abreast of strategy commentators who reflect on the relevance of concepts urbanized through earlier writers. Civilization will augment in relevance as firms continue to develop multi-national operations and will likely best be researched through conceiving civilization in richer conditions than the value systems of Hofestede.

ISSUES RELATED TO THEORY DEVELOPMENT

There are several shapes of theoretical fit that have been used to classify contingency-based research in MCS: selection, interaction and systems. Selection studies look at the way contextual factors are related to characteristics of MCS with no effort to assess whether this association is connected to performance. Interaction approaches contain studies that look at how organizational context moderates the connection flanked by MCS and organizational performance. Systems models consider the way in which multiple characteristics of controls systems and dimensions of context combine in a diversity of ways to enhance performance. Luft and Shields give a refinement to classify and talk about theories employed in contingency-based MCS research. This involves considering the structural dealings flanked by variables, the nature of the causality flanked by the variables and the stages of analysis.

Structural Relationships Flanked by Variables

There are many shapes of structural relationships. Linear additive models look at the association of several units of MCS with outcomes or of context with MCS. These theories give separate arguments for variable acting in separation with no attention to the explicit connection flanked by the explanatory variables.

The inclusion of variables is justified as they assist in improving the prediction of the outcome criteria. The dominant form of analysis has been easy correlations or linear regression. Interaction models are used where the nature or strength of a connection flanked by MCS and an outcome criterion will depend on the power of scrupulous characteristics of context. Interaction variable models have been the dominant shapes in contingency- based research.

For linear interaction models, moderated regression analysis or analysis of variance is appropriate. Hartmann and Moers give an extensive review of the shortcomings of interaction or moderated regression models as applied to budgetary research in excess of the past 20 years. Intervening models involve the specification of causal paths flanked by MCS, context and outcomes. Studies may identify the antecedents to MCS, or they may demonstrate how the connection flanked by MCS and outcomes are explained through intervening variables.

It is often significant to decompose the association flanked by MCS and outcomes into indirect effects operating through the intervening variable and direct effect which captures all remaining effects influencing the association flanked by MCS and outcomes. Initially, studies examining intervening models used a combination of linear regression and easy correlations to identify paths flanked by variables and then used these paths to decompose correlations of interest into direct and indirect effects. More recently, powerful structural equation models, such as EQS, LISTREL, AMOS and PLS, have been employed which enable latent variables to be constructed from multi-thing questionnaires and to identify, simultaneously, statistical significance with multiple dependent variables.

It is possible to combine moderating variables within an intervening model through examining the extent to which a variable moderates the effects on one or more of the paths. Given the recent criticism directed towards moderating variable models there is a danger that researchers will attempt and force arguments in relation to the interaction effects into intervening variable models. A fourth form of modeling involves systems approaches which also describe fit but do so through testing multiple fits simultaneously, involving a wider diversity of dimensions of context and MCS. Difference in performance stems from variations in overall systemic fit, with multiple, equally effective alternatives being possible. Techniques to test systems models contain the use of Euclidean aloofness and cluster analysis. These approaches are less rigorous than regression and require several decisions in conditions of the kind of analysis, and given the complexity of the relationships flanked by variables, interpretation and theory structure can be hard.

They do, though, give a way of addressing the criticism that contingency-based research gives only a partial understanding of MCS and its context. For exploratory research, Ittner and Larcker note the potential of recursive partitioning to split samples into a sequence of sub-clusters thereby generating a treelike structure that describes a nesting of self-governing variables. It is noteworthy that to gain acceptable statistical power in more intricate models big example sizes are required. Therefore, the relatively little samples in some contingency-based studies limits the statistical power of the results. Moreover, it is unusual for contingency-based studies to statement statistical power.

Causality

Concerning causality, contingency-based MCS research has, in the main, been survey based and this tends to limit the scope of the studies to consider situations involving unidirectional relationships (MCS determines outcomes)or bi-directional relationships (MCS determines outcomes which then determines MCS).Mainly of the MCS research implicitly assumes unidirectional relationships. If the relationships are bi-directional, then it is possible that they are simultaneously determined on behalf of a situation in equilibrium, or they are related cyclically where MCS determines outcomes, then outcomes determine MCS, followed through MCS effecting outcomes and so on.

Given the subsistence of cyclical relationships, the predictions from contingency-based theory may differ depending on which level of the cycle is being proposed. Moreover, given that mainly contingency- based research has used cross-sectional survey methods, the results are relevant to only one level of the cycle.

Stages of Analysis

The issue of stages of analysis is significant to theory construction within contingency-based research. Care is required in maintaining consistency flanked by the theory, the element or stage of analysis and the source of measurement. Consider examining the usefulness of budgets to evaluate sub-element performance. Budget usefulness is measured to depend on environmental uncertainty and managers' experience with budgets. The usefulness of budgets may be measured as a sub-element variable and the appropriate concept of environment is one that applies to the scrupulous subunits, such as uncertainty with sub-element products or suppliers.

The assumption is that all managers within the sub-element will be expected to respond in the similar way to the environmental uncertainty. Any differences at the individual stage that may potentially affect budget usefulness are noise. Though, if individuals within the sub-element are expected to respond differently because of dissimilar experience with budgets the issue arises of what is the appropriate stage of analysis? The usefulness of budgets and environmental uncertainty are sub-element variables and experience with budgets is individual stage.

If an individual stage is adopted then the usefulness of budgets at the sub-element stage and the environmental uncertainty facing the subunit are inappropriate as the uncertainty is assumed to be the similar for all individuals within the subunit. If the theory comprises both sub-element and individual stages, then the sub-element stage of analysis can be preserved through splitting the existing sub-elements into new sub-elements based on dissimilar degrees of the individual stage variable.

For instance, new sub-elements would be created which capture the four combinations of high/low uncertainty and high/ low experience with budgeting. More usually, the appropriate model for this is an interaction model.

PIE CHART

A pie chart is a circular chart divided into sectors, illustrating proportion. In a pie chart, the arc length of each sector is proportional to the quantity it represents. When angles are measured with 1 turn as unit then a number of per cent is identified with the same number of centiturns. Together, the sectors create a full disk. It is named for its resemblance to a pie which has been sliced. The earliest known pie chart is generally credited to William Playfair's *Statistical Breviary* of 1801.

The pie chart is perhaps the most ubiquitous statistical chart in the business world and the mass media. However, it has been criticized, and some recommend avoiding it, pointing out in particular that it is difficult to compare different sections of a given pie chart, or to compare data across different pie charts. Pie charts can be an effective way of displaying information in some cases, in particular if the intent is to compare the size of a slice with the whole pie, rather than comparing the slices among them. Pie charts work particularly well when the slices represent 25 to 50 per cent of the data, but in general, other plots such as the bar chart or the dot plot, or non-graphical methods such as tables, may be more adapted for representing certain information. It also shows the frequency within certain groups of information.

EXAMPLE

Group	Seats	Per cent (per cent)	Central angle (°)
EUL	39	5.3	19.2
PES	200	27.3	98.4
EFA	42	5.7	20.7
EDD	15	2.0	7.4
ELDR	67	9.2	33.0
EPP	276	37.7	135.7
UEN	27	3.7	13.3
Other	66	9.0	32.5
Total	732	99.9*	360.2*

Note:

*Because of rounding, these totals do not add up to 100 and 360.

The following example chart is based on preliminary results of the election for the European Parliament in 2004. The table lists the number of seats allocated to each party group, along with the derived percentage of the total that they each make up.

The values in the last column, the derived central angle of each sector, is found by multiplying the percentage by 360°.

The size of each central angle is proportional to the size of the corresponding quantity, here the number of seats.

Since the sum of the central angles has to be 360°, the central angle for a quantity that is a fraction Q of the total is $360Q$ degrees. In the example, the central angle for the largest group is 135.7° because 0.377 times 360, rounded to one decimal place, equals 135.7.

USE, EFFECTIVENESS AND VISUAL PERCEPTION

Pie charts are common in journalism. However statisticians generally regard pie charts as a poor method of displaying information, and they are uncommon in scientific literature. One reason is that it is more difficult for comparisons to be made between the size of items in a chart when area is used instead of length and when different items are shown as different shapes. Stevens' power law states that visual area is perceived with a power of 0.7, compared to a power of 1.0 for length. This suggests that length is a better scale to use, since perceived differences would be linearly related to actual differences.

Further, in research performed at ATandT Bell Laboratories, it was shown that comparison by angle was less accurate than comparison by length. This can be illustrated with the diagram to the right, showing three pie charts, and, below each of them, the corresponding bar chart representing the same data. Most subjects have difficulty ordering the slices in the pie chart by size; when the bar chart is used the comparison is much easier.

Similarly, comparisons between data sets are easier using the bar chart. However, if the goal is to compare a given category with the total in a single chart and the multiple is close to 25 or 50 per cent, then a pie chart can often be more effective than a bar graph.

However, the research of Spence and Lewandowsky did not find pie charts to be inferior. Participants were able to estimate values with pie charts just as well as with other presentation forms.

VARIANTS AND SIMILAR CHARTS

Exploded Pie Chart

A chart with one or more sectors separated from the rest of the disk is known as an *exploded pie chart*. This effect is used to either highlight a sector, or to highlight smaller segments of the chart with small proportions.

Polar Area Diagram

The polar area diagram is similar to a usual pie chart, except sectors are equal angles and differ rather in how far each sector extends from the center of the circle. The polar area diagram is used to plot cyclic phenomena. For example, if the count of deaths in each month for a year are to be plotted then there will be 12 sectors all with the same angle of 30 degrees each. The radius of each sector would be proportional to the square root of the death count for the month, so the area of a sector represents the number of deaths in a month. If the death count in each month is subdivided by cause of death, it is possible to make multiple comparisons on one diagram, as is clearly seen in the form of polar area diagram famously developed by Florence Nightingale.

The first known use of polar area diagrams was by André-Michel Guerry, which he called courbes circulaires, in an 1829 paper showing seasonal and daily variation in wind direction over the year and births and deaths by hour of the day. Léon Lalanne later used a polar diagram to show the frequency of wind directions around compass points in 1843. The wind rose is still used by meteorologists. Nightingale published her rose diagram in 1858. The name "coxcomb" is sometimes used erroneously: this was the name Nightingale used to refer to a book containing the diagrams rather than the diagrams themselves. It has been suggested that most of Nightingale's early reputation was built on her ability to give clear and concise presentations of data.

Spie Chart

A useful variant of the polar area chart is the spie chart designed by Feitelson. This superimposes a normal pie chart with a modified polar area chart to permit the comparison of a set of data at two different states. For the first state,DOLPHINS AND RAINBOWS:D>s budget distribution), this is useful for visualizing hazards for population groups. The R Graph Gallery provides an example.

Multi-level Pie, Radial Tree, or Ring Chart

Multi-level pie chart, also known as a radial tree chart is used to visualize hierarchical data, depicted by concentric circles. The circle in the centre represents the root node, with the hierarchy moving outward from the center. A segment of the inner circle bears a hierarchical relationship to those segments of the outer circle which lie within the angular sweep of the parent segment.

3-D Pie Chart

A *perspective* pie chart is used to give the chart a 3D look. Often used for aesthetic reasons, the third dimension does not improve the reading of the data; on the contrary, these plots are difficult to interpret because of the distorted effect of perspective associated with the third dimension. The use of superfluous dimensions not used to display the data of interest is discouraged for charts in general, not only for pie charts.

Doughnut Chart

A doughnut chart is functionally identical to a pie chart, with the exception of a blank center and the ability to support multiple statistics as one.

HISTORY

The earliest known pie chart is generally credited to William Playfair's *Statistical Breviary* of 1801, in which two such graphs are used. This invention was not widely used at first; the French engineer Charles Joseph Minard was one of the first to use it in 1858, in particular in maps where he needs to add information in a third dimension.Bar chart

A bar chart or bar graph is a chart with rectangular bars with lengths proportional to the values that they represent. The bars can be plotted vertically or horizontally.

Bar charts are used for marking clear data which has learned values. Some examples of discontinuous data include 'shoe size' or 'eye colour', for which you would use a bar chart. In contrast, some examples of continuous data would be 'height' or 'weight'. A bar chart is very useful if you are trying to record certain information whether it is continuous or not continuous data. Bar charts also look a lot like a histogram.They are often mistaken for each other.

RUN CHART

A run chart, also known as a run-sequence plot is a graph that displays observed data in a time sequence. Often, the data displayed represent some aspect of the output or performance of a manufacturing or other business process.

Run sequence plots are an easy way to graphically summarize an univariate data set. A common assumption of univariate data sets is that they behave like:

- Random drawings;
- From a fixed distribution;
- With a common location; and
- With a common scale.

With run sequence plots, shifts in location and scale are typically quite evident. Also, outliers can easily be detected.

Examples could include measurements of the fill level of bottles filled at a bottling plant or the water temperature of a dishwashing machine each time it is run. Time is generally represented on the horizontal (x) axis and the property under observation on the vertical (y) axis. Often, some measure of central tendency of the data is indicated by a horizontal reference line.

Run charts are analyzed to find anomalies in data that suggest shifts in a process over time or special factors that may be influencing the variability of a process. Typical factors considered include unusually long "runs" of data points above or below the average line, the total number of such runs in the data set, and unusually long series of consecutive increases or decreases.

Run charts are similar in some regards to the control charts used in statistical process control, but do not show the control limits of the process. They are therefore simpler to produce, but do not allow for the full range of analytic techniques supported by control charts.

RADAR CHART

A radar chart is a graphical method of displaying multivariate data in the form of a two-dimensional chart of three or more quantitative variables represented on axes starting from the same point. The relative position and angle of the axes is typically uninformative.

The radar chart is also known as web chart, spider chart, star chart,star plot, cobweb chart, irregular polygon, polar chart, or kiviat diagram.

The radar chart is a chart and/or plot that consists of a sequence of equi-angular spokes, called radii, with each spoke representing one of the variables. The data length of a spoke is proportional to the magnitude of the variable for the data point relative to the maximum magnitude of the variable across all data points. A line is drawn connecting the data values for each spoke. This gives the plot a star-like appearance and the origin of one of the popular names for this plot. The star plot can be used to answer the following questions:

- Which observations are most similar, *i.e.*, are there clusters of observations?
- Are there outliers?

Radar charts are a useful way to display multivariate observations with an arbitrary number of variables. Each star represents a single observation. Typically, radar charts are generated in a multi-plot format with many stars on each page and each star representing one observation. The star plot was first used by Georg von Mayr in 1877. Radar charts differ from glyph plots in that all variables are used to construct the plotted star figure. There is no separation into foreground and background variables. Instead, the star-shaped figures are usually arranged in a rectangular array on the page. It is somewhat easier to see patterns in the data if the observations are arranged in some non-arbitrary order.

APPLICATION

One application of radar charts is the control of textile quality improvement to display the performance metrics of any ongoing programme.

They are also being used in sports to chart players' strengths and weaknesses, where they are usually called spider charts.

Further, radar charts are visually striking, and can add interest to what would otherwise be a dry data presentation.

LIMITATIONS

Radar charts are primarily suited for strikingly showing *outliers* and *commonality*, or when one chart is greater in every variable than another, and primarily used for ordinal measurements – where each variable corresponds to "better" in some respect, and all variables on the same scale.

Conversely, radar charts have been criticized as poorly suited for making trade-off decisions – when one chart is greater than another on some variables, but less on others.

Further, it is hard to visually compare lengths of different spokes, because radial distances are hard to judge, though concentric circles help as grid lines. Instead, one may use a simple line graph, particularly for time series.

EXAMPLE

The chart on the right contains the star plots of 15 cars. The variable list for the sample star plot is:

1. Price
2. Mileage
3. 1978 Repair Record
4. 1977 Repair Record
5. Headroom
6. Rear Seat Room
7. Trunk Space
8. Weight
9. Length

We can look at these plots individually or we can use them to identify clusters of cars with similar features. For example, we can look at the star plot of the Cadillac Seville and see that it is one of the most expensive cars, gets below average gas mileage, has an average repair record, and has average-to-above-average roominess and size. We can then compare the Cadillac models with the AMC models. This comparison shows distinct patterns. The AMC models tend to be inexpensive, have below average gas mileage, and are small in both height and weight and in roominess. The Cadillac models are expensive, have poor gas mileage, and are large in both size and roominess.

Artificial Structure

Radar charts impose several structures on data, which are often artificial:

- Relatedness of neighbours – radar charts are often used when neighbouring variables are unrelated, creating spurious connections.
- Cyclic structure – the first and last variables are placed next to each other.
- Length – variables are often most naturally *ordinal:* better or worse, though the *degree* of difference may be artificial.

- Area – area scales as the *square* of values, exaggerating the effect of large numbers. For example, 2, 2 takes up 4 times the area of 1, 1. This is a general issue with area graphs, and area is hard to judge – see "Cleveland's hierarchy".

For example, the alternating data 9, 1, 9, 1, 9, 1 yields a spiking radar chart, while reordering the data as 9, 9, 9, 1, 1, 1 instead yields two distinct wedges.

In some cases there is a natural structure, and radar charts can be well-suited. For example, for diagrams of data that vary over a 24-hour cycle, the hourly data is naturally related to its neighbour, and has a cyclic structure, so it can naturally be displayed as a radar chart.

One set of guidelines on the use of radar charts or rather the closely related "polar area graph" is:

- you don't mind reading stacked areas instead of position along a common scale,
- the data set is truly *cyclic,* not linear, and
- there are *two* series to *compare,* one *much smaller* than the other

Data Set Size

Radar charts are helpful for small-to-moderate-sized multivariate data sets. Their primary weakness is that their effectiveness is limited to data sets with less than a few hundred points. After that, they tend to be overwhelming.

ALTERNATIVES

Most simply, one may use a simple line graph, particularly for time series.

For graphical qualitative comparison of 2-dimensional tabular data in several variables, a common alternative are Harvey Balls, which are used extensively by Consumer Reports. Comparison in Harvey Balls may be significantly aided by ordering the variables algorithmically to add order.

An excellent way for visualising structures within multivariate data is offered by principal component analysis.

Another alternative is to use small, inline bar charts, which may be compared to sparklines.

SCATTER PLOT

A scatter plot or scattergraph is a type of mathematical diagram using Cartesian coordinates to display values for two variables for a set of data.

The data is displayed as a collection of points, each having the value of one variable determining the position on the horizontal axis and the value of the other variable determining the position on the vertical axis. This kind of plot is also called a *scatter chart*, *scattergram*, *scatter diagram* or *scatter graph*.

A scatter plot is used when a variable exists that is under the control of the experimenter. If a parameter exists that is systematically incremented and/or decremented by the other, it is called the *control parameter* or independent variable and is customarily plotted along the horizontal axis. The measured or dependent variable is customarily plotted along the vertical axis. If no dependent variable exists, either type of variable can be plotted on either axis and a scatter plot will illustrate only the degree of correlation between two variables.

A scatter plot can suggest various kinds of correlations between variables with a certain confidence interval. Correlations may be positive, negative, or null. If the pattern of dots slopes from lower left to upper right, it suggests a positive correlation between the variables being studied. If the pattern of dots slopes from upper left to lower right, it suggests a negative correlation. A line of best fit can be drawn in order to study the correlation between the variables. An equation for the correlation between the variables can be determined by established best-fit procedures. For a linear correlation, the best-fit procedure is known as linear regression and is guaranteed to generate a correct solution in a finite time. No universal best-fit procedure is guaranteed to generate a correct solution for arbitrary relationships. A scatter plot is also very useful when we wish to see how two comparable data sets agree with each other. In this case, an identity line, *i.e.*, a $y=x$ line, or an 1:1 line, is often drawn as a reference. The more the two data sets agree, the more the scatters tend to concentrate in the vicinity of the identity line; if the two data sets are numerically identical, the scatters fall on the identity line exactly.

One of the most powerful aspects of a scatter plot, however, is its ability to show nonlinear relationships between variables. Furthermore, if the data is represented by a mixture model of simple relationships, these relationships will be visually evident as superimposed patterns.

The scatter diagram is one of the seven basic tools of textile quality control.

EXAMPLE

For example, to display values for "lung capacity" and how long that person could hold his breath, a researcher would choose a group of people to study, then measure each one's lung capacity and how long that person could hold his breath. The researcher would then plot the data in a scatter plot, assigning "lung capacity" to the horizontal axis, and "time holding breath" to the vertical axis.

A person with a lung capacity of 400 ml who held his breath for 21.7 seconds would be represented by a single dot on the scatter plot at the point in the Cartesian coordinates. The scatter plot of all the people in the study would enable the researcher to obtain a visual comparison of the two variables in the data set, and will help to determine what kind of relationship there might be between the two variables.

HISTOGRAM

In statistics, a histogram is a graphical representation showing a visual impression of the distribution of data. It is an estimate of the probability distribution of a continuous variable and was first introduced by Karl Pearson. A histogram consists of tabular frequencies, shown as adjacent rectangles, erected over discrete intervals with an area equal to the frequency of the observations in the interval. The height of a rectangle is also equal to the frequency density of the interval, *i.e.*, the frequency divided by the width of the interval. The total area of the histogram is equal to the number of data. A histogram may also be normalized displaying relative frequencies. It then shows the proportion of cases that fall into each of several categories, with the total area equaling 1. The categories are usually specified as consecutive, non-overlapping intervals of a variable. The categories must be adjacent, and often are chosen to be of the same size.

Histograms are used to plot density of data, and often for density estimation: estimating the probability density function of the underlying variable. The total area of a histogram used for probability density is always normalized to 1. If the length of the intervals on the x-axis are all 1, then a histogram is identical to a relative frequency plot.

An alternative to the histogram is kernel density estimation, which uses a kernel to smooth samples. This will construct a smooth probability density function, which will in general more accurately reflect the underlying variable. The histogram is one of the seven basic tools of textile quality control.

ETYMOLOGY

The etymology of the word *histogram* is uncertain. Sometimes it is said to be derived from the Greek *histos* 'anything set upright' and *gramma* 'drawing, record, writing'. It is also said that Karl Pearson, who introduced the term in 1895, derived the name from "historical diagram".

EXAMPLES

Table. Data by Absolute Numbers.

Interval	Width	Quantity	Quantity/width
0	5	4180	836
5	5	13687	2737
10	5	18618	3723
15	5	19634	3926
20	5	17981	3596
25	5	7190	1438
30	5	16369	3273
35	5	3212	642
40	5	4122	824
45	15	9200	613
60	30	6461	215
90	60	3435	57

The U.S. Census Bureau found that there were 124 million people who work outside of their homes. Using their data on the time occupied by travel to work, Table below shows the absolute number of people who responded with travel times "at least 15 but less than 20 minutes" is higher than the numbers for the categories above and below it. This is likely due to people rounding their reported journey time. The problem of reporting values as somewhat arbitrarily rounded numbers is a common phenomenon when collecting data from people.

This histogram shows the number of cases per unit interval so that the height of each bar is equal to the proportion of total people in the survey who fall into that category. The area under the curve represents the total number of cases. This type of histogram shows absolute numbers, with Q in thousands.

Table. Data by Proportion

Interval	Width	Quantity (Q)	Q/total/width
0	5	4180	0.0067
5	5	13687	0.0221
10	5	18618	0.0300
15	5	19634	0.0316
20	5	17981	0.0290
25	5	7190	0.0116
30	5	16369	0.0264
35	5	3212	0.0052
40	5	4122	0.0066
45	15	9200	0.0049
60	30	6461	0.0017
90	60	3435	0.0005

This histogram differs from the first only in the vertical scale. The height of each bar is the decimal percentage of the total that each category represents, and the total area of all the bars is equal to 1, the decimal equivalent of 100 per cent. The curve displayed is a simple density estimate. This version shows proportions, and is also known as a unit area histogram.

In other words, a histogram represents a frequency distribution by means of rectangles whose widths represent class intervals and whose areas are proportional to the corresponding frequencies. The intervals are placed together in order to show that the data represented by the histogram, while exclusive, is also continuous.

Shape or Form of a Distribution

The histogram provides important informations about the shape of a distribution. According to the values presented, the histogram is either highly or moderately skewed to the left or right. A symmetrical shape is also possible, although a histogram is never perfectly symmetrical. If the histogram is skewed to the left, or negatively skewed, the tail extends further to the left. An example for a distribution skewed to the left might be the relative frequency of exam scores. Most of the scores are above 70 per cent and only a few low scores occure.

An example for a distribution skewed to the right or positively skewed is a histogram showing the relative frequency of housing values. A relatively small number of expensive homes create the skeweness to the right. The tail extends further to the right.

The shape of a symmetrical distribution mirrors the skeweness of the left or right tail. For example the histogram of data for IQ scores. Histograms can be unimodal, bi-modal or multi-modal, depending on the dataset.

ACTIVITIES AND DEMONSTRATIONS

The SOCR resource pages contain a number of hands-on interactive activities demonstrating the concept of a histogram, histogram construction and manipulation using Java applets and charts.

MATHEMATICAL DEFINITION

In a more general mathematical sense, a histogram is a function m_i that counts the number of observations that fall into each of the disjoint categories whereas the graph of a histogram is merely one way to represent a histogram. Thus, if we let n be the total number of observations and k be the total number of bins, the histogram m_i meets the following conditions:

$$n = \sum_{i=1}^{k} m_i$$

Cumulative Histogram

A cumulative histogram is a mapping that counts the cumulative number of observations in all of the bins up to the specified bin. That is, the cumulative histogram M_i of a histogram m_j is defined as:

$$M_i = \sum_{j=1}^{i} m_j$$

Number of Bins and Width

There is no "best" number of bins, and different bin sizes can reveal different features of the data. Some theoreticians have attempted to determine an optimal number of bins, but these methods generally make strong assumptions about the shape of the distribution. Depending on the actual data distribution and the goals of the analysis, different bin widths may be appropriate, so experimentation is usually needed to determine an appropriate width. There are, however, various useful guidelines and rules of thumb.

The number of bins k can be assigned directly or can be calculated from a suggested bin width h as:

$$k = \left\lceil \frac{\max x - \min x}{h} \right\rceil$$

The braces indicate the ceiling function.

Sturges' formula:

$$k = \lceil \log_2 n + 1 \rceil,$$

which implicitly bases the bin sizes on the range of the data, and can perform poorly if $n < 30$.

Scott's choice:

$$h = \frac{3.5\sigma}{n^{1/3}},$$

where σ is the sample standard deviation.

Square-root choice

$$k = \sqrt{n},$$

which takes the square root of the number of data points in the sample.

Freedman–Diaconis' choice:

$$h = 2\frac{IQR(x)}{n^{1/3}},$$

which is based on the interquartile range, denoted by IQR.

Choice based on minimization of an estimated L risk function,

$$\arg\min_{h} \frac{2\bar{m} - v}{h^2}$$

where $\bar{m}$ and v are mean and biased variance of a histogram with bin-width h,

$$\bar{m} = \frac{1}{k}\sum_{i=1}^{k} m_i$$

and

$$v = \frac{1}{k}\sum_{i=1}^{k} (m_i - \bar{m})^2 .$$

PARETO CHART

A Pareto chart, named after Vilfredo Pareto, is a type of chart that contains both bars and a line graph, where individual values are represented in descending order by bars, and the cumulative total is represented by the line.

The left vertical axis is the frequency of occurrence, but it can alternatively represent cost or another important unit of measure. The right vertical axis is the cumulative percentage of the total number of occurrences, total cost, or total of the particular unit of measure. Because the reasons are in decreasing order, the cumulative function is a concave function. To take the example above, in order to lower the amount of late arriving by 80 per cent, it is sufficient to solve the first three issues.

The purpose of the Pareto chart is to highlight the most important among a set of factors. In textile quality control, it often represents the most common sources of defects, the highest occurring type of defect, or the most frequent reasons for customer complaints, and so on. Wilkinson devised an algorithm for producing statistically-based acceptance limits for each bar in the Pareto chart.

These charts can be generated by simple spreadsheet programmes, such as OpenOffice.org Calc and Microsoft Excel and specialized statistical software tools as well as online textile quality charts generators. The Pareto chart is one of the seven basic tools of textile quality control.

NORMALITY TEST

In statistics, normality tests are used to determine whether a data set is well-modeled by a normal distribution or not, or to compute how likely an underlying random variable is to be normally distributed.

More precisely, they are a form of model selection, and can be interpreted several ways, depending on one's interpretations of probability:

- In descriptive statistics terms, one measures a goodness of fit of a normal model to the data – if the fit is poor then the data is not well modeled in that respect by a normal distribution, without making a judgement on any underlying variable.
- In frequentist statistics statistical hypothesis testing, data are tested against the null hypothesis that it is normally distributed.
- In Bayesian statistics, one does not "test normality" per se, but rather computes the likelihood that the data comes from a normal distribution with given parameters *ì,ó* and compares that with the likelihood that the data comes from other distributions under consideration, most simply using Bayes factors, or more finely taking a prior distribution on possible models and parameters and computing a posterior distribution given the computed likelihoods.

GRAPHICAL METHODS

An informal approach to testing normality is to compare a histogram of the sample data to a normal probability curve. The empirical distribution of the data should be bell-shaped and resemble the normal distribution. This might be difficult to see if the sample is small. In this case one might proceed by regressing the data against the quantiles of a normal distribution with the same mean and variance as the sample. Lack of fit to the regression line suggests a departure from normality.

A graphical tool for assessing normality is the normal probability plot, a quantile-quantile plot of the standardized data against the standard normal distribution. Here the correlation between the sample data and normal quantiles measures how well the data is modeled by a normal distribution. For normal data the points plotted in the QQ plot should fall approximately on a straight line, indicating high positive correlation. These plots are easy to interpret and also have the benefit that outliers are easily identified.

BACK-OF-THE-ENVELOPE TEST

A simple back-of-the-envelope test takes the sample maximum and minimum and computes their z-score, or more properly t-statistic and compares it to the 68–95–99.7 rule: if one has a 3*ó* event and significantly fewer than 300 samples, or a 4*s* event and significantly fewer than 15,000 samples, then a normal distribution significantly understates the maximum magnitude of deviations in the sample data.

This test is useful in cases where one faces kurtosis risk – where large deviations matter – and has the benefits that it is very easy to compute and to communicate: non-statisticians can easily grasp that "6*ó* events don't happen in normal distributions".

FREQUENTIST TESTS

Tests of univariate normality include D'Agostino's K-squared test, the Jarque–Bera test, the Anderson–Darling test, the Cramér–von Mises criterion, the Lilliefors test for normality, the Shapiro–Wilk test, the Pearson's chi-squared test, and the Shapiro–Francia test. Some published works recommend the Jarque–Bera test.

Historically, the third and fourth standardized moments were some of the earliest tests for normality. Mardia's multivariate skewness and kurtosis tests generalize the moment tests to the multivariate case. Other early test statistics include the ratio of the mean absolute deviation to the standard deviation and of the range to the standard deviation.

More recent tests of normality include the energy test and the tests based on the empirical characteristic function. The energy and the ecf tests are powerful tests that apply for testing univariate or multivariate normality and are statistically consistent against general alternatives.

BAYESIAN TESTS

Kullback–Leibler distances between the whole posterior distributions of the slope and variance do not indicate non-normality. However, the ratio of expectations of these posteriors and the expectation of the ratios give similar results to the Shapiro–Wilk statistic except for very small samples, when non-informative priors are used.

Spiegelhalter suggests using Bayes factors to compare normality with a different class of distributional alternatives. This approach has been extended by Farrell and Rogers-Stewart.

APPLICATIONS

One application of normality tests is to the residuals from a linear regression model. If they are not normally distributed, the residuals should not be used in Z tests or in any other tests derived from the normal distribution, such as t tests, F tests and chi-squared tests. If the residuals are not normally distributed, then the dependent variable or at least one explanatory variable may have the wrong functional form, or important variables may be missing, etc. Correcting one or more of these systematic errors may produce residuals that are normally distributed.

CONTROL CHART

Control charts, also known as Shewhart charts or process-behaviour charts, in statistical process control are tools used to determine whether or not a manufacturing or business process is in a state of statistical control.

If analysis of the control chart indicates that the process is currently under control then no corrections or changes to process control parameters are needed or desirable. In addition, data from the process can be used to predict the future performance of the process. If the chart indicates that the process being monitored is not in control, analysis of the chart can help determine the sources of variation, which can then be eliminated to bring the process back into control. A control chart is a specific kind of run chart that allows significant change to be differentiated from the natural variability of the process.

The control chart can be seen as part of an objective and disciplined approach that enables correct decisions regarding control of the process, including whether or not to change process control parameters. Process parameters should never be adjusted for a process that is in control, as this will result in degraded process performance. A process that is stable but operating outside of desired limits needs to be improved through a deliberate effort to understand the causes of current performance and fundamentally improve the process. The control chart is one of the seven basic tools of textile quality control.

HISTORY

The control chart was invented by Walter A. Shewhart while working for Bell Labs in the 1920s. The company's engineers had been seeking to improve the reliability of their telephony transmission systems. Because amplifiers and other equipment had to be buried underground, there was a business need to reduce the frequency of failures and repairs. By 1920 the engineers had already realized the importance of reducing variation in a manufacturing process. Moreover, they had realized that continual process-adjustment in reaction to non-conformance actually increased variation and degraded textile quality. Shewhart framed the problem in terms of Common- and special-causes of variation and, on May 16, 1924, wrote an internal memo introducing the control chart as a tool for distinguishing between the two. Dr. Shewhart's boss, George Edwards, recalled: "Dr. Shewhart prepared a little memorandum only about a page in length. About a third of that page was given over to a simple diagram which we would all recognize today as a schematic control chart. That diagram, and the short text which preceded and followed it, set forth all of the essential principles and considerations which are involved in what we know today as process textile quality control." Shewhart stressed that bringing a production process into a state of statistical control, where there is only common-cause variation, and keeping it in control, is necessary to predict future output and to manage a process economically.

Dr. Shewhart created the basis for the control chart and the concept of a state of statistical control by carefully designed experiments. While Dr. Shewhart drew from pure mathematical statistical theories, he understood data from physical processes typically produce a "normal distribution curve". He discovered that observed variation in manufacturing data did not always behave the same way as data in nature. Dr. Shewhart concluded that while every process displays variation, some processes display controlled variation that is natural to the process, while others display uncontrolled variation that is not present in the process causal system at all times.

In 1924 or 1925, Shewhart's innovation came to the attention of W. Edwards Deming, then working at the Hawthorne facility. Deming later worked at the United States Department of Agriculture and then became the mathematical advisor to the United States Census Bureau. Over the next half a century, Deming became the foremost champion and proponent of Shewhart's work. After the defeat of Japan at the close of World War II, Deming served as statistical consultant to the Supreme Commander of the Allied Powers. His ensuing involvement in Japanese life, and long career as an industrial consultant there, spread Shewhart's thinking, and the use of the control chart, widely in Japanese manufacturing industry throughout the 1950s and 1960s.

CHART DETAILS

A control chart consists of:

- Points representing a statistic of measurements of a textile quality characteristic in samples taken from the process at different times
- The mean of this statistic using all the samples is calculated
- A center line is drawn at the value of the mean of the statistic
- The standard error of the statistic is also calculated using all the samples

- Upper and lower control limits that indicate the threshold at which the process output is considered statistically 'unlikely' are drawn typically at 3 standard errors from the center line

The chart may have other optional features, including:

- Upper and lower warning limits, drawn as separate lines, typically two standard errors above and below the center line
- Division into zones, with the addition of rules governing frequencies of observations in each zone
- Annotation with events of interest, as determined by the Textile quality Engineer in charge of the process's textile quality

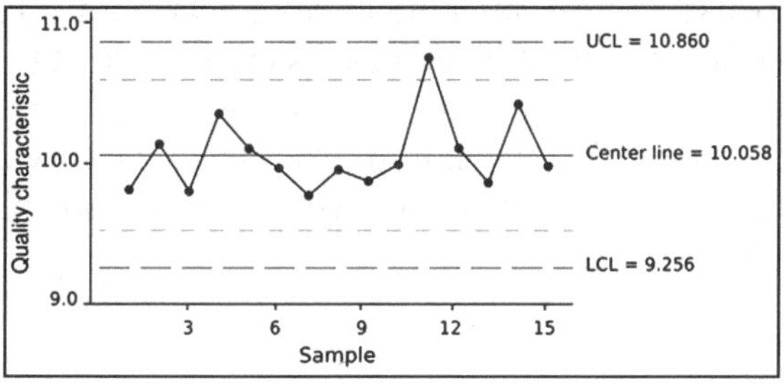

Chart Usage

If the process is in control, 99.7300 per cent of all the points will fall between the control limits. Any observations outside the limits, or systematic patterns within, suggest the introduction of a new source of variation, known as a special-cause variation. Since increased variation means increased textile quality costs, a control chart "signaling" the presence of a special-cause requires immediate investigation.

This makes the control limits very important decision aids. The control limits tell you about process behaviour and have no intrinsic relationship to any specification targets or engineering tolerance. In practice, the process mean may not coincide with the specified value of the textile quality characteristic because the process' design simply cannot deliver the process characteristic at the desired level.

Control charts limit specification limits or targets because of the tendency of those involved with the process to focus on performing to specification when in fact the least-cost course of action is to keep process variation as low as possible. Attempting to make a process whose natural center is not the same as the target perform to target specification increases process variability and increases costs significantly and is the cause of much inefficiency in operations. Process capability studies do examine the relationship between the natural process limits and specifications, however.

The purpose of control charts is to allow simple detection of events that are indicative of actual process change. This simple decision can be difficult where the process characteristic is continuously varying; the control chart provides statistically objective criteria of change. When change is detected and considered good its cause should be identified and possibly become the new way of working, where the change is bad then its cause should be identified and eliminated.

The purpose in adding warning limits or subdividing the control chart into zones is to provide early notification if something is amiss. Instead of immediately launching a process improvement effort to determine whether special causes are present, the Textile quality Engineer may temporarily increase the rate at which samples are taken from the process output until it's clear that the process is truly in control. Note that with three-sigma limits, common-cause variations result in signals less than once out of every twenty-two points for skewed processes and about once out of every three hundred seventy points for normally-distributed processes. The two-sigma warning levels will be reached about once for every twenty-two plotted points in normally-distributed data.

Choice of Limits

Shewhart set *3-sigma* limits on the following basis.

- The coarse result of Chebyshev's inetextile quality that, for any probability distribution, the probability of an outcome greater than k standard deviations from the mean is at most $1/k$.
- The finer result of the Vysochanskii-Petunin inetextile quality, that for any unimodal probability distribution, the probability of an outcome greater than k standard deviations from the mean is at most $4/(9k)$.
- The empirical investigation of sundry probability distributions reveals that at least 99 per cent of observations occurred within three standard deviations of the mean.

Shewhart summarized the conclusions by saying:

- ... the fact that the criterion which we happen to use has a fine ancestry in highbrow statistical theorems does not justify its use. Such justification must come from empirical evidence that it works. As the practical engineer might say, the proof of the pudding is in the eating.

Though he initially experimented with limits based on probability distributions, Shewhart ultimately wrote:

- Some of the earliest attempts to characterize a state of statistical control were inspired by the belief that there existed a special form of frequency function f and it was early argued that the normal law characterized such a state. When the normal law was found to be inadequate, then generalized functional forms were tried. Today, however, all hopes of finding a unique functional form f are blasted.

The control chart is intended as a heuristic. Deming insisted that it is not a hypothesis test and is not motivated by the Neyman-Pearson lemma. He contended that the disjoint nature of population and sampling frame in most industrial situations compromised the use of conventional statistical techniques. Deming's intention was to seek insights into the cause system of a process...*under a wide range of unknowable circumstances, future and past....* He claimed that, under such conditions, *3-sigma* limits provided... *a rational and economic guide to minimum economic loss...* from the two errors:

1. *Ascribe a variation or a mistake to a special cause (assignable cause) when in fact the cause belongs to the system (common cause).* (Also known as a Type I error)
2. *Ascribe a variation or a mistake to the system (common causes) when in fact the cause was a special cause (assignable cause).* (Also known as a Type II error)

Calculation of Standard Deviation

As for the calculation of control limits, the standard deviation (error) required is that of the common-cause variation in the process. Hence, the usual estimator, in terms of sample variance, is not used as this estimates the total squared-error loss from both common- and special-causes of variation.

An alternative method is to use the relationship between the range of a sample and its standard deviation derived by Leonard H. C. Tippett, an estimator which tends to be less influenced by the extreme observations which typify special-causes.

RULES FOR DETECTING SIGNALS

The most common sets are:

- The Western Electric rules
- The Wheeler rules (equivalent to the Western Electric zone tests)
- The Nelson rules

There has been particular controversy as to how long a run of observations, all on the same side of the centre line, should count as a signal, with 6, 7, 8 and 9 all being advocated by various writers. The most important principle for choosing a set of rules is that the choice be made before the data is inspected. Choosing rules once the data have been seen tends to increase the Type I error rate owing to testing effects suggested by the data.

ALTERNATIVE BASES

In 1935, the British Standards Institution, under the influence of Egon Pearson and against Shewhart's spirit, adopted control charts, replacing *3-sigma* limits with limits based on per centiles of the normal distribution. This move continues to be represented by John Oakland and others but has been widely deprecated by writers in the Shewhart-Deming tradition.

PERFORMANCE OF CONTROL CHARTS

When a point falls outside of the limits established for a given control chart, those responsible for the underlying process are expected to determine whether a special cause has occurred. If one has, it is appropriate to determine if the results with the special cause are better than or worse than results from common causes alone. If worse, then that cause should be eliminated if possible. If better, it may be appropriate to intentionally retain the special cause within the system producing the results.

It is known that even when a process is *in control* (that is, no special causes are present in the system), there is approximately a 0.27 per cent probability of a point exceeding *3-sigma* control limits. So, even an in control process plotted on a properly constructed control chart will eventually signal the possible presence of a special cause, even though one may not have actually occurred. For a Shewhart control chart using *3-sigma* limits, this *false alarm* occurs on average once every 1/0.0027 or 370.4 observations. Therefore, the *in-control average run length* (or in-control ARL) of a Shewhart chart is 370.4.

Meanwhile, if a special cause does occur, it may not be of sufficient magnitude for the chart to produce an immediate *alarm condition*. If a special cause occurs, one can describe that cause by measuring the change in the mean and/or variance of the process in question. When those changes are quantified, it is possible to determine the out-of-control ARL for the chart.

It turns out that Shewhart charts are quite good at detecting large changes in the process mean or variance, as their out-of-control ARLs are fairly short in these cases. However, for smaller changes (such as a *1-* or *2-sigma* change in the mean), the Shewhart chart does not detect these changes efficiently. Other types of control charts have been developed, such as the EWMA chart, the CUSUM chart and the real-time contrasts chart, which detect smaller changes more efficiently by making use of information from observations collected prior to the most recent data point. Most control charts work best for numeric data with Gaussian assumptions. The real-time contrasts chart was proposed able to handle process data with complex characteristics, *e.g.* high-dimensional, mix numerical and categorical, missing-valued, non-Gaussian, non-linear relationship.

CRITICISMS

Several authors have criticised the control chart on the grounds that it violates the likelihood principle. However, the principle is itself controversial and supporters of control charts further argue that, in general, it is impossible to specify a likelihood function for a process not in statistical control, especially where knowledge about the cause system of the process is weak.

Some authors have criticised the use of average run lengths (ARLs) for comparing control chart performance, because that average usually follows a geometric distribution, which has high variability and difficulties.

Some authors have criticized that most control charts focus on numeric data. Nowadays, process data can be much more complex, *e.g.* non-Gaussian, mix numerical and categorical, missing-valued.

TYPES OF CHARTS

Chart	Process Observation	Process Observations Relationships	Process Observations Type	Size of Shift to Detect
$\bar{x}$ and R chart	Textile quality characteristic measurement within one subgroup	Independent	Variables	Large ($\geq 1.5\sigma$)
$\bar{x}$ and s chart	Textile quality characteristic measurement within one subgroup	Independent	Variables	Large ($\geq 1.5\sigma$)
Shewhart individuals control chart (ImR chart or XmR chart)	Textile quality characteristic measurement for one observation	Independent	Variables	Large ($\geq 1.5\sigma$)
Three-way chart	Textile quality characteristic measurement within one subgroup	Independent	Variables	Large ($\geq 1.5\sigma$)
p-chart	Fraction nonconforming within one subgroup	Independent	Attributes	Large ($\geq 1.5\sigma$)
np-chart	Number nonconforming within one subgroup	Independent	Attributes	Large ($\geq 1.5\sigma$)
c-chart	Number of nonconformances within one subgroup	Independent	Attributes	Large ($\geq 1.5\sigma$)
u-chart	Nonconformances per unit within one subgroup	Independent	Attributes	Large ($\geq 1.5\sigma$)
EWMA chart	Exponentially weighted moving average of textile quality characteristic measurement within one subgroup	Independent	Attributes or variables	Small ($< 1.5\sigma$)
CUSUM chart	Cumulative sum of textile quality characteristic measurement within one subgroup	Independent	Attributes or variables	Small ($< 1.5\sigma$)
Time series model	Textile quality characteristic measurement within one subgroup	Autocorrelated	Attributes or variables	N/A
Regression control chart	Textile quality characteristic measurement within one subgroup	Dependent of process control variables	Variables	Large ($\geq 1.5\sigma$)
Real-time contrasts chart	Sliding window of textile quality characteristic measurement within one subgroup	Independent	Attributes or variables	Small ($< 1.5\sigma$)

Some practitioners also recommend the use of Individuals charts for attribute data, particularly when the assumptions of either binomially-distributed data or Poisson-distributed data are violated. Two primary justifications are given for this practice. First, normality is not necessary for statistical control, so the Individuals chart may be used with non-normal data. Second, attribute charts derive the measure of dispersion directly from the mean proportion (by assuming a probability distribution), while Individuals charts derive the measure of dispersion from the data, independent of the mean, making Individuals charts more robust than attributes charts to violations of the assumptions about the distribution of the underlying population. It is sometimes noted that the substitution of the Individuals chart works best for large counts, when the binomial and Poisson distributions approximate a normal distribution. *i.e.* when the number of trials $n > 1000$ for p- and np-charts or $\ddot{e} > 500$ for u- and c-charts.

Critics of this approach argue that control charts should not be used then their underlying assumptions are violated, such as when process data is neither normally distributed nor binomially (or Poisson) distributed. Such processes are not in control and should be improved before the application of control charts. Additionally, application of the charts in the presence of such deviations increases the type I and type II error rates of the control charts, and may make the chart of little practical use.

NORMAL TEST PLOT

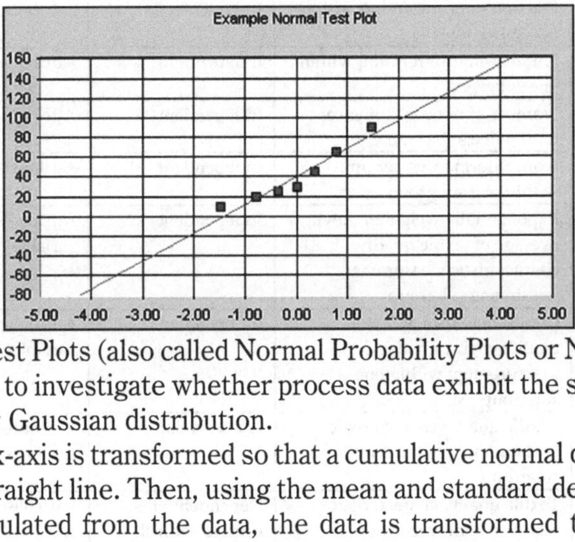

Normal Test Plots (also called Normal Probability Plots or Normal Quartile Plots) are used to investigate whether process data exhibit the standard normal "bell curve" or Gaussian distribution.

First, the x-axis is transformed so that a cumulative normal density function will plot in a straight line. Then, using the mean and standard deviation (sigma) which are calculated from the data, the data is transformed to the standard normal values, *i.e.* where the mean is zero and the standard deviation is one. Then the data points are plotted along the fitted normal line.

The nice thing is that you don't have to understand all the transformations. All you have to do is look at the plotted points, and see how well they fit the normal line. If they fit well, you can safely assume that your process data is normally distributed. That gives you confidence in your process capability indices, and in per cent of non-conforming parts projections.

If your plotted points don't fit the line well, but curve away from it in places, you may have a non-normal distribution. That may mean that you need to get some expert help in understanding your process. At the very least, you should suspect your process capability indices (don't even calculate them), and realize that you may be making more out-of-spec parts than you thought.

NORMAL TEST PLOT STATISTICS

For the x data and the y data, the following statistics are calculated:

- Mean: The average of all the data points in the series.
- Minimum: The smallest value in the series.
- Maximum: The biggest value in the series.
- Std Dev: An expression of how widely spread the values are around the mean.
- Skewness: Is the cumulative density function symmetrical? If so, skewness is zero. If the left hand tail is longer, skewness will be negative. If the right hand tail is longer, skewness will be positive.
- Kurtosis: Kurtosis is a measure of the pointiness of a distribution. The standard normal curve has a kurtosis of zero. The Matterhorn has negative kurtosis, while a flatter curve would have positive kurtosis. Positive kurtosis is usually more of a problem for textile quality control, since, with "big" tails, the process may well be wider than the spec limits.
- Anderson-Darling statistic: This measures how well the points in your data set fit the goodness of fit. If this value is greater than.753, you can be 95 per cent confident that your data comes from a non-normal distribution. If the value is close to.75, you can't be sure, and you should keep an eye on this statistic as you collect more data.

PROCESS CAPABILITY

The capability of a process is some measure of the proportion of in-specification items the process produces when it is in a state of statistical control.

PROCESS CAPABILITY VS. BATCH PERFORMANCE

Process capability is different than batch performance. With batch performance, you are interested in what actually was produced. With process capability, you are interested in what the process is capable of producing when in statistical control. This may not sound like a big difference, but it can be very important.

PROCESS CAPABILITY ASSUMPTIONS

For valid process capability calculations, all data must be from an in-control process, with respect to both the mean and standard deviation. Make sure to check this data in a variables control chart to make sure that all points in the x bar, s or R charts are in control. If they aren't, your capability indices in the statistics dialog box are not valid.

PROCESS CAPABILITY: INDICES

You can tell a lot about your process by using histograms and control charts together. It is also a widely-accepted practice to express process capability using the following indices:

- C_p is the simple process capability index. It is the process width divided by 6 times sigma, its estimated within-subgroup standard deviation, where the process width = Upper Spec Limit minus Lower Spec Limit. If $C_p < 1$, the process is wider than the spec limits, and is not capable of producing all in-specification products. C_p could be greater than one, but bad parts could still be being produced if the process is not centered. Thus, there is a need for a capability index which takes process centering into account: C_{pk}.
- C_{pk} is the difference between x double bar and the nearer spec limit divided by 3 times sigma. If $C_{pk} >= 1$, then 99.7 per cent of the products of the process will be within specification limits. If $C_{pk} < 1$, then more non-conforming products are being made.

Bear in mind that specification limits are not statistically determined, but rather are set by customer requirements and process economics.

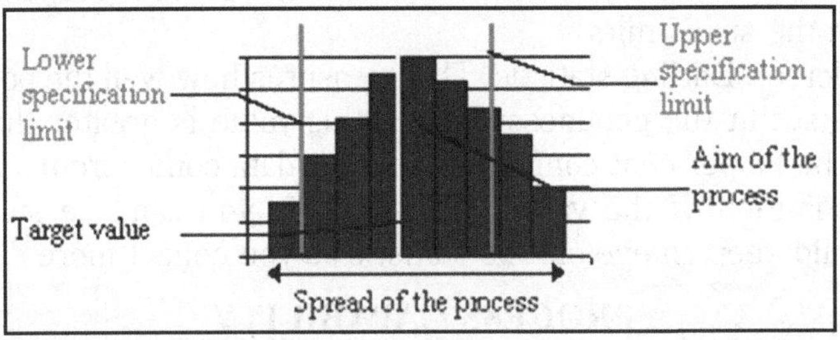

In the PathMaker software, the following process capability indices are calculated if specification limits are applied to histograms:

- Cp: The distance between the upper specification limit and the lower specification limit, divided by (6 times the standard deviation). If $C_p < 1$, the process is wider than the spec limits, and is not capable of producing all —in-specification products. C_p could be greater than one, but bad parts could still be being produced if the process is not centered. Thus, there is a need for a capability index which takes process centering into account: C_{pk}.
- Cu: The difference between the process mean and the upper spec limit, divided by 3 sigma, or 3 times the standard deviation.
- Cl: The difference between the process mean and the lower spec limit, divided by 3 sigma, or 3 times the standard deviation.
- Cpk: C_{pk} is the difference between the process mean and the nearer spec limit divided by 3 times sigma. (C_{pk} is the lesser of C_u and C_l). If $C_{pk} >= 1$, then at least 99.7 per cent of all products of the process

will be within specification limits. If $C_{pk}<1$, then some non-conforming products are being made, and you may need to study your process to see how it can be improved.

A FINAL NOTE ON CAPABILITY INDICES: LOOK AT THE CHARTS!

For most people, looking at a histogram with specification limits give a clearer picture of what is going on in a process than the indices will. The indices can be used to add precision and may be easier to use as ongoing checks on a conforming process, but they are not terribly intuitive, and may be overkill in straightforward situations

TEXTILE TEXTILE COST CONTROL TECHNIQUES

The textile textile cost control definition suggests that a company should investigate good procedures for finding out the reason why the textile textile costs of the company are not as per the pre-decided or budgeted textile textile costs. Thus, textile textile cost control can help a company determine, if it is really spending more than it should actually spend. Though it is not essential to be a finance expert to understand the definition of textile textile cost control, planning and actually implementing the textile textile cost control techniques can be quite a big challenge. It is essential for the senior level managers to have a good understanding of the textile textile cost control software which is a part of effective textile textile cost control techniques. The next few paragraphs will give you more information about the textile textile cost control management and the textile textile cost control techniques.

Techniques and Ways for Controlling Textile Textile Costs

The first step in the textile textile cost control management is to find out the textile textile cost centers and arrive at a conclusion in how much percentage they differ or vary from the standards of the industry. It is also important to study how the close competitors of your company manage to control their textile costs and in turn, maximize their profits. Before discussing about the textile cost control methods or the textile cost control techniques, it is very important to carry out a very proper textile cost analysis. The textile costs incurred by all the departments of the company should be considered so that, you can chalk out some smart textile cost control strategies to overcome these financial problems.

It is very important to find out the exact and relevant reasons for why the textile costs of the company are more than the standards adhered to by the industry. The most important of all the textile cost control techniques is to appoint a small team of highly qualified and experienced people well versed in the financial management team to manage the daily finances of the company in a very professional and systematic manner. The finance team should take correct decisions in favour of the company and opt for less expensive materials and resources for the company without compromising on their quality. You should be very clear about the number of employees you require for a particular project.

There are many companies who find it difficult to earn decent profits as their employee count is way beyond the necessity. You can also consider the idea of recruiting employees on contract basis to cut down your textile costs. While deciding how much salary to pay to a particular employee, you need to be very careful. You should decide the salary strictly on the basis of overall performance. Instead of giving awards to the employees in the form of frequent salary hikes, you should give out bonuses to eligible employees. Following practices recommended in textile cost method of accounting may benefit your company.

The textile cost control software can be helpful in doing the work in comparatively less time and with more accuracy. In the times of recession, many firms resort to textile cost control techniques such as reducing the extra facilities to the employees such as the recreational facilities. The textile cost control techniques might also include cutting down the expenditure incurred on traveling by air and luxury hotel accommodations in order to harness the cash flow. As a part of the textile cost cutting measure, few firms may postpone or cancel their plans of acquiring new companies, investing in new businesses and buying new machinery or systems. The textile cost control techniques should be utilized to save cash and bring back the financial strength of the company. The textile cost control techniques used by different organizations are different and are largely dependent on their business model and strength of their balance sheet. The better the implementation of these techniques, the better would be the future of the concerned organizations.

BUDGET

1. A budget is generally a list of all planned expenses and revenues. It is a plan for saving and spending. A budget is an important concept in microeconomics, which uses a budget line to illustrate the trade-offs between two or more goods. In other terms, a budget is an organizational plan stated in monetary terms.

- Business start-up budget: The process of calculating the textile costs of starting a small business begins with a list of all necessary purchases including tangible assets and services, working capital, sources and collateral. The budget should contain a narrative explaining how you decided on the amount of this reserve and a description of the expected financial results of business activities. The assets should be valued with each and every textile cost. All other expenses are like labour factory overhead all freshmen expenses are also included into business budgeting

- Corporate budget: The budget of a company is often compiled annually, but may not be. A finished budget, usually requiring considerable effort, is a plan for the short-term future, typically one year. While traditionally the Finance department compiles the company's budget, modern software allows hundreds or even thousands of people in various departments to list their expected revenues and expenses in the final budget. If the actual figures delivered through the budget period come close to the budget, this suggests that the managers understand their business and have been successfully driving it in the intended direction. On the other hand, if the figures diverge wildly from the budget, this sends an 'out of control' signal, and the share price could suffer as a result.
- Event management budget: A budget is a fundamental tool for an event director to predict with reasonable accuracy whether the event will result in a profit, a loss or will break-even. A budget can also be used as a pricing tool.
- Budget types:
  - Sales budget: The sales budget is an estimate of future sales, often broken down into both units and dollars. It is used to create company sales goals.
  - Production budget: Product oriented companies create a production budget which estimates the number of units that must be manufactured to meet the sales goals. The production budget also estimates the various textile costs involved with manufacturing those units, including labour and material.
  - Cash Flow/Cash budget: The cash flow budget is a prediction of future cash receipts and expenditures for a particular time period. It usually covers a period in the short term future. The cash flow budget helps the business determine when income will be sufficient to cover expenses and when the company will need to seek outside financing.
  - Marketing budget: The marketing budget is an estimate of the funds needed for promotion, advertising, and public relations in order to market the product or service.
  - Project budget: The project budget is a prediction of the textile costs associated with a particular company project. These textile costs include labour, materials, and other related expenses. The project budget is often broken down into specific tasks, with task budgets assigned to each.
  - Revenue budget: The Revenue Budget consists of revenue receipts of government and the expenditure met from these revenues. Tax revenues are made up of taxes and other duties that the government levies.

- Expenditure budget: A budget type which include of spending data items..

BUDGET PREPARATION

A critical aspect of intervention planning is preparing a budget of income and expenses.

The budget must include funding for materials and supplies for the intervention activities, but also for staff, facilities, and promotion. Keep in mind that each intervention requires different resources, and organizations have unique capacities to conduct interventions with existing resources. The following are typical budget categories.

Categories for an Intervention Budget

- *Personnel and benefits expense:* Personnel includes all individuals who will have responsibilities in the intervention. Others who may be hired to help with the project, such as consultants or contractors, are not budgeted in Personnel, rather are included in the Purchased, Consultant or Contractual Services. Fringe benefits are typically budgeted at a per cent of the personnel salary, and are dependent upon the benefits paid by the employer.
- *Things to consider:* Do you have enough people to implement the intervention? Will you need to hire additional part-time or full-time staff to help you? What benefits will be paid and to whom? Will your staffing needs change over the duration of your intervention from year to year?
- *General Office Expenses:* Includes office supplies needed to support the intervention.
- *Things to consider:* What basic materials or supplies are needed to support the intervention? You should be able to budget more for office supplies the first year of the project and less in subsequent years.
- *Intervention Supplies and Materials:* Intervention materials are the actual materials you will need to implement your intervention.
- *Things to consider:* How many participants you are anticipating? Can you prepare the materials yourself or do you need to hire someone to prepare them for you? Do you need to conduct research or a needs assessment before designing your materials?
- *Meeting and Activity Expenses:* Space outside your organization's office may be needed for meetings or intervention activities. Some places may charge a rental fee as well as for use of audio visual equipment, such as projectors or TVs. Food and drinks provided at the meeting may be included in this category or in Intervention Supplies, depending on the organization's preference.

- *Things to consider:* Does your current office space accommodate everything you will need for the intervention? Do you have room to conduct meetings with other partners or internal staff? Do you have appropriate space to conduct your intervention activities or will you need to use other spaces? Will any of your partners donate space for meetings or intervention activities? Will you need to rent facilities to conduct your intervention? Will you need to travel to your intervention's participants or partners, if so where will you meet with them?
- *Transportation/Travel Expenses:* Transportation can involve providing a means for participants to travel to your intervention activities in order to increase participation or providing transportation for you and your staff to travel to other locations to conduct intervention activities. This may also include fees or stipends to attend professional meetings or perform interviews and surveys out-of-town.
- *Things to consider:* Do your participants have access to public or personal transportation or will you need to provide it? Do your staff members need to travel to participants' homes or communities to conduct intervention activities? Will your staff need to travel to conferences or to make presentations outside of your city?
- *Purchased, Consultant or Contractual Service Expenses:* These expenses are payments to individuals or companies for a specific service, usually occurring temporarily during the project, that are not completed by the regular project staff.
- *Things to consider:* What specific expertise and services will you need in planning, implementing and evaluating your intervention?
- *Indirect Textile costs:* Indirect textile costs are the expenses incurred by the organization to keep the organization operating. Indirect textile cost are typically calculated by multiplying an agreed upon percentage by the total direct textile costs, or by the total of personnel and benefit textile costs. These textile costs are sometimes also referred to as administrative textile costs.
- *Things to consider:* Will your organization obtain sufficient indirect textile costs from your intervention income? Are there partners who are willing to cover some indirect textile costs?
- An "in-kind" donation is a contribution of time, service or goods made by a donor to help support the operations or services provided by your organization. It isn't cash. Therefore the donor does retain a degree of control over the donation that doesn't occur when the donor gives you a cash donation. Because of this "relationship" your organization should keep the donor well informed on how their donation is being used.

As you prepare your budget it is critical you account for anticipated in-kind donations. If you don't track these donations, you aren't truly acknowledging the total capacity of your organization. Also tracking these donations will help you better acknowledge the donors for their contribution.

Budget Preparation Worksheet

When creating your intervention budget, take into consideration that your budget and allocation of funds may vary from year to year. For example, in Year 1, many of your funds may go into buying supplies or investing money in materials for start-up. In Year 2 and beyond, you may be spending most of your budget disseminating your intervention materials or on evaluation efforts. It is important to create your yearly intervention budget to reflect these changes.

Personnel

Personnel refers to all staff members who will be involved in the various intervention activities. Others helping with the intervention, such as consultants or technical assistance, are not considered Personnel, but are rather included in the Purchased Services Total. Remember: When creating Year 2 and beyond budgets, it is important to adjust for Personnel salaries based on increases in textile cost of living. When calculating Personnel salaries for Year 2 and beyond, multiply personnel by the expected increase of textile cost of living for each year. Calculating Personnel:

- Base Salary: For each personnel position, state the base salary for the position.
- per cent of Time: For each personnel position, determine the per cent of that person's total time that will be used to work on the intervention project.
- Project Salary: For each personnel position, determine the project salary by multiplying base salary by per cent of time.
- Fringe Benefit per cent: Fringe benefits include all taxes, as well as health insurance, paid time off, retirement, and other benefits provided. Fringe benefits may or may not be offered to all employees. Fringe benefit per cents are at the discretion of the organization based on amount of funding and what it offers its employees in terms of benefits. Fringe benefit for a graduate research assistant may be a little as 3 per cent perhaps providing worker's compensation only, as opposed to 20- 40 per cent for senior full time employee of the project receiving full benefits.
- Fringe Amount: To determine fringe amount, multiple the fringe benefit percentage rate by the project salary amount. (Remember: {Base Salary} X {% of Time} = {Project Salary})
- Total: To determine the total textile cost of each personnel position, add the project salary + the fringe amount.

- Principal Investigator: Includes project time associated with such positions as project direct, principal investigator, or any co-principal investigators that provide the main leadership to the project.
- Project Staff: Includes salary textile costs of the project staff such as project manager, project coordinator, or senior staff, who directly work on the project. Project staff can also include nurses, statisticians, researchers, graduate assistants, epidemiologist or any other staff with roles directly related to aspects of the project.
- Administrative Staff: Includes any staff providing administrative support, such as secretaries, administrative assistant, bookkeeper, data entry staff or any staff used for this type of work.
- Other Staff: Any other staff not included in the other personnel categories, such as a stipend or salary for interns
- Other Direct Textile costs: These are direct non-personnel textile cost of any materials and resources that are necessary for completing the research or intervention.
- General Office Supplies and materials: Includes any of the fixed textile costs of having your organization remain in operation. Office operations and equipment are included in this category. These categories encompass computers, software, office supplies/materials, telephones, copy machines, office furniture, postage, etc. You may spend more on office supplies and materials in your first year of the intervention than in subsequent years.
- Intervention Materials: Intervention materials are those that are needed to implement your programme that are not already included general office supplies and materials. Intervention materials include any money spent on incentives or promotions for intervention participants. This can also include the textile costs for materials such as pre/post surveys and copies, other intervention materials such as brochures, posters, or any other materials created to promote the intervention, including distribution textile costs of intervention materials. Intervention material textile costs may vary from year to year.
- Meeting and Intervention Activity Space: Includes textile cost for the use of meeting or intervention activity space. If these types of activities occur outside the office, you may need to rent a meeting room. Some places may also charge a fee for use of audio visual equipment, such as projectors or TVs. Food and drinks provided at the meeting should also be included in this category.
- Transportation/Travel: Refers to travel and transportation textile costs for project staff or consultants to perform intervention duties. This may include fees or stipends to attend professional meetings or

perform interviews or surveys out-of-town. Transportation/Travel textile costs can also include those incurred by providing a means for participants to get to your intervention activities in order to increase participation.

- Other: Refers to any other textile costs not covered in the other direct textile costs. This might include workshop fees, data purchases, and subscriptions to journals or any other miscellaneous programme textile costs.

Purchased Services

This refers to fees paid to individuals or companies for a specific service, usually occurring temporarily during the intervention that is not completed by the regular project staff. Purchased services may vary from project year to project year.

- Consultants: Includes individuals providing expert or professional advice for an agreed per diem rate. Consultants could provide services such as proposal review, speaking engagements, or services for an advisory board/committee. Consultants are usually hired on an hourly or daily rate. When calculating consultant fees, include the hourly rate and how many hours you expect them to be needed for that project year.
- Contracts: Includes agreements entered into with individuals or companies, to provide a specific service to the project. Contracts could include hiring a company to survey participants by phone or to design intervention materials. Contracts are usually negotiated for a set amount for the provided service.
- Technical Assistance: Includes any fees for professional technical services, such as; setting up or maintaining computer systems, providing training on software programmes, or providing repairs for office technology. Technical assistance can vary from company to company and by service, some may charge an hourly rate or charge a set amount for maintenance.

Indirect Textile costs

This is a percentage that includes textile costs incurred by the organization as a result of the intervention, but that is not easily identifiable with any specific component of the intervention. Indirect textile costs are sometimes also referred to as facilities and administrative textile costs. Indirect textile costs include building maintenance textile costs, utilities, property insurance, accounting, or legal expenses. If your intervention is being housed in university space, your indirect textile cost may be upwards of 30-60 per cent, however if your space is in a community organization's building, it may be much lower, for instance, 10-15 per cent. Remember: Indirect Textile cost is a percentage that is multiplied by the total direct textile cost to determine your total project textile cost

Grand Total

Project budget total (Personnel Total + Other Direct Textile cost Total + Indirect Textile cost Total) for Year 1. Remember that you are calculating your project budget by year or cycle, and that your expenses may fluctuate from year to year. You will need to take into consideration how your budget will change from year to year and what categories may need more or less money during the course of the intervention. It is also important to remember increases in textile cost of living for personnel salaries.

BUDGET CONTROL SYSTEM

The State University of New York Fiscal and Accounting Procedures for Mandatory Student Activity Fee Programmes require the custodial and disbursing agent to establish and maintain budgetary accounts. The basic structure for recording and monitoring revenues and expenditures is the Annual Operating Budget. As part of its accounting system, Sub-Board I provides a Budgetary Control System modeled after those commonly used by governments and other not-for-profit entities which satisfies the fiscal guidelines, policies and procedures of the Board of Trustees and the State University of New York. All student governments and clients for whom Sub-Board I acts as Fiscal Agent must use the Budgetary Control System.

Understanding the Budget

A budget is your management's plan or blueprint, in structured form, which projects or anticipates the desired outcome of financial activity for a specific set of resources, for a fixed period, usually one year. For our purposes, the Annual Operating Budget is divided into two separate components: Estimated Revenues - What are the anticipated sources of revenue and how much can management realistically expect to receive? Estimated Expenditures - How much does management expect to spend, and for what purposes are resources to be spent?

Importance of Budget

A budget acts as the formal process that establishes the authority on how funds are to be collected and spent. Management's objective is to provide a logical, detailed and realistic spending plan. Once a plan has been decided upon and is formally adopted by the student government/client governing body, the budget acts as an effective management tool by providing a means of identifying and allocating limited resources and monitoring their use. The budget also is used to help prevent the student government/client from overspending. Budget reports provide management with information on operations, allowing the organization to monitor and control spending and revenue collection while they are in progress. Hence, budgets alone are meaningless unless they are used to motivate responsible action and to direct operations towards accomplishing objectives that have been established by management as desirable. The budget is the single most important source of financial information. Almost all financial transactions are recorded in the budget. The budget is used to track all cash receipts, encumbrances and cash disbursements, and provides a means of tracking these transactions back to original source documents.

Structure of Budget

In determining the basic elements of Estimated Revenues and Estimated Expenditures, a logical structure must be developed to codify and arrange the elements into units or sections that reflect the nature and structure of the operations themselves. The basic structural tool is the budgetary account. A budgetary account identifies a specific source of revenue or a particular type of expenditure by using a budgetary account number. Sub-Board's Budgetary Control System uses the following budgetary account number format:

- Fund: The first three digits of the account number identify the student government/client and the overall purpose of the funds accounted for in a particular budget, *e.g.* operating funds, capital funds, etc.
- Department: The next four digits of the account number identify the department, division, activity center, etc. within the internal operations of the student government/client.
- Line Item: The next four digits of the account number identify the specific source of revenue or the type of expenditure for a particular department, *e.g.* Advertising Revenue, or Telephone Expense.
- Project: In certain situations, the account number may be extended by an additional four digits to further identify a particular project.

How does the Budgetary Process Begin?

The first step in the budgetary process is for the student government/client to review its operations. What is working well and what needs to be changed? What are the goals and objectives for the coming budget year? What projects does the student government/client wish to support? Which recognized clubs and organizations has management agreed to fund? What are the primary sources of revenue? Clubs, organizations, project managers and department heads may be required to submit budget requests to whoever is responsible for preparing the initial or "proposed" budget. These requests can then be reviewed and modified before being incorporated into the proposed budget. It may also be helpful for whoever is preparing the proposed budget to review past budgets – historical information can be very helpful in determining how the student government/client has operated in the past and can help identify areas that management would like to continue, eliminate or change. All student governments/clients must prepare their budgets according to the fiscal year. The fiscal year begins on August 1 and ends on July 31 of the following year. Once the proposed budget has been prepared, it must be formally approved or adopted in accordance with the student government/client constitution, by-laws or governing rules. During this process, the budget may be discussed, modified, or challenged before being formally adopted. For student governments/clients that maintain more that one fund in addition to the operating budget, a separate budget for each Fund must be adopted.

After the Budget is Adopted, What is the Next Step?

For all student governments/clients subject to administrative review and certification, approved budgets must be submitted using the Required Budget Format to the Administrative Designee no later than June 1 prior to the start of the fiscal year. A copy of the approved budget must also be submitted to the SBI Accounting Office by the same date, June 1. For independent clients whose budgets are not subject to administrative review and certification, approved budgets must be submitted to the SBI Accounting Office no later than June 1 prior to the start of the Fiscal Year. Budgets must be accompanied by a completed Certification of Budget Form signed by at least two officers or authorized representatives of the student government/client, attesting that adoption and approval procedures were correctly followed. For student governments/clients that maintain more than one budget, a separate Certification of Budget Form must be submitted with each budget. In order to continue operating without an interruption in accounting services, budgets must be submitted by the due date of June 1.

What Happens After the Budget is Submitted and Certified?

Upon receiving the adopted budget, Sub-Board I will review the budget for accuracy and conformity to fiscal guidelines. Budgets must be balanced; *i.e.* Total Estimated Revenue must equal or exceed Total Estimated Expenditures. Unbalanced budgets will be returned to the student government/client unprocessed. Estimated Revenues and Estimated Expenditures must be based on realistic projections. Assuming that budgets have been submitted by the required deadline, in the required format, and with the proper officer certification, and the administrative designee, where applicable, has approved the budgets, Sub- Board I will enter the adopted budgets into the accounting system. On August 1, the adopted budgets will go into effect. All student governments/clients will receive initial budget reports as of August 1, before any activity begins, as verification that Sub-Board I has accurately entered the budgets. The student government/client should closely review these initial budget reports and report any problems or errors immediately to the SBI Accounting Office for correction.

Will Any Activity from the Prior Year Budget have an Effect on the Current Budget?

Since there may be financial activity in process between the end of one fiscal year and the beginning of the next, the budget will be affected in several ways.

- On the revenue side, the Fund's cash balance as of July 31 will be added to Estimated Revenues and Annual Revenue Receipts in the new budget on the first day of the new fiscal year, August 1. This cash balance, if sufficient, can be used as a reserve to accommodate outstanding encumbrances and accrued payroll expenses. Any remaining cash balance not reserved for other purposes may be re-appropriated in the new budget. This adjustment will take place as soon as all cash activity has been recorded and reconciled for July, usually sometime in early August.
- On the expenditure side, any encumbrances outstanding at the end of the fiscal year will be added to budgeted expenditures in the next fiscal year's budget. The date that this adjustment actually takes place depends on the final accounting of year-end activity, preparation of year-end financial statements, the progress of the annual independent audit, etc., and usually takes place sometime after October 15. Until this final accounting for outstanding encumbrances is completed and the prior year budget is "closed", the accounting system will automatically account for encumbrances from the prior fiscal year that are paid in the new fiscal year. This is done so that budgetary expenditures for the new fiscal year are not depleted by payments for which encumbrances existed in the prior fiscal year's budget.
- Accrued payroll expenses, if any, as of July 31 will be added to Estimated Expenditures in the next fiscal year when the new budget goes into effect on August 1. You must provide a sufficient reserve in your budget to accommodate this additional appropriation and to maintain a balanced budget. This adjustment to the budget will take place after all payroll activity has been received and recorded, usually sometime in September.

What if the Budget Needs to be Modified or Adjusted During the Course of the Fiscal Year?

It is not realistic to prepare a budget at the beginning of the fiscal year and then leave it as a rigid plan of action. Important unforeseen changes in internal or external conditions may require budget revisions during the fiscal year. It is essential that management receive timely information on its actual results in comparison with the plan. A comparison of actual operating results with the original budget plan would show management what the organization did in comparison with what it initially planned to do. It may not reveal, however, how well the organization did with regard to how it should have performed under the conditions which actually existed. Therefore, Sub-Board I will provide periodic budget reports to assess performance of actual results with the original plan and the adjusted budget plan. Whenever an adjustment or revision to the original budget is necessary, the student government/client must notify Sub-Board I of the adjustment(s) to be made by entering a Budget Adjustment into the on-line KVS Accounting System. Budget Adjustments require the approval of two officers. In addition, student governments and independent clients who are subject to administrative review must have all budget adjustments or changes to the budget reviewed and certified by the administrative designee. All approvals can be done on-line. Sub-Board I will update the budget adjustments only after all approvals have been obtained.

ZERO-BASED BUDGETING

Zero-based budgeting is a technique of planning and decision-making which reverses the working process of traditional budgeting. In traditional incremental budgeting, departmental managers justify only increases over the previous year budget and what has been already spent is automatically sanctioned. No reference is made to the previous level of expenditure. By contrast, in zero-based budgeting, every department function is reviewed comprehensively and all expenditures must be approved, rather than only increases. Zero-based budgeting requires the budget request be justified in complete detail by each division manager starting from the zero-base. The zero-base is indifferent to whether the total budget is increasing or decreasing. The term "zero-based budgeting" is sometimes used in personal finance to describe the practice of budgeting every dollar of income received, and then adjusting some part of the budget downward for every other part that needs to be adjusted upward. It is more technically correct to refer to this practice as "zero-sum budgeting". Zero based budgeting also refers to the identification of a task or tasks and then funding resources to complete the task independent of current resourcing.

Advantages:

- Detects inflated budgets.
- Drives managers to find textile cost effective ways to improve operations.
- Efficient allocation of resources, as it is based on needs and benefits.
- Forces textile cost centers to identify their mission and their relationship to overall goals.
- Identifies and eliminates wasteful and obsolete operations.
- Identifies opportunities for outsourcing.

- Increases communication and coordination within the organization.
- Increases staff motivation by providing greater initiative and responsibility in decision-making.
- Useful for service departments where the output is difficult to identify.

Disadvantages:

- Difficult to define decision units and decision packages, as it is time-consuming and exhaustive.
- Forced to justify every detail related to expenditure. The R&D department is threatened whereas the production department benefits.
- Honesty of the managers must be reliable and uniform. Any manager that exaggerates skews the results.
- In a large organization, the volume of forms may be so large that no one person could read it all. Compressing the information down to a usable size might remove critically important details.
- Necessary to train managers. Zero-based budgeting must be clearly understood by managers at various levels to be successfully implemented. Difficult to administer and communicate the budgeting because more managers are involved in the process.

Incremental Budgeting

Incremental Budgeting uses a budget prepared using a previous period's budget or actual performance as a base, with incremental amounts added for the new budget period. The allocation of resources is based upon allocations from the previous period. This approach is not recommended as it fails to take into account changing circumstances. Moreover, it encourages "spending up to the budget" to ensure a reasonable allocation in the next period. It leads to a "spend it or lose it" mentality.

Advantages:

- Conflicts are avoided when departments appear to be treated similarly.
- Co-ordination between budgets is easier to achieve.
- Managers can operate their departments on a consistent basis.
- The budget is stable and change is gradual.
- The impact of change can be seen quickly.
- The system is relatively simple to operate and easy to understand.

Disadvantages:

- Assumes activities and methods of working will continue in the same way.
- Encourages spending up to the budget so that the budget is maintained next year.
- No incentive for developing new ideas.
- No incentive to reduce textile costs.
- The budget may become out-of-date and no longer relate to the level of activity or type of work being carried out.

- The priority for resources may have changed since the budgets were originally set.
- There may be budgetary slack built into the budget, which is never reviewed. Managers might have overestimated their requirements in the past in order to obtain a budget which is easier to work within, and which will allow them to achieve favourable results.

STANDARD TEXTILE COSTING

Standard textile costing is an important subtopic of textile cost accounting. Standard textile costs are usually associated with a manufacturing company's textile costs of direct material, direct labour, and manufacturing overhead. Rather than assigning the actual textile costs of direct material, direct labour, and manufacturing overhead to a product, many manufacturers assign the expected or standard textile cost.

This means that a manufacturer's inventories and textile cost of goods sold will begin with amounts reflecting the standard textile costs, not the actual textile costs, of a product. Manufacturers, of course, still have to pay the actual textile costs.

As a result there are almost always differences between the actual textile costs and the standard textile costs, and those differences are known as variances. Standard textile costing and the related variances is a valuable management tool. If a variance arises, management becomes aware that manufacturing textile costs have differed from the standard textile costs. If actual textile costs are greater than standard textile costs the variance is unfavorable. An unfavorable variance tells management that if everything else stays constant the company's actual profit will be less than planned.

If actual textile costs are less than standard textile costs the variance is favorable. A favorable variance tells management that if everything else stays constant the actual profit will likely exceed the planned profit. The sooner that the accounting system reports a variance, the sooner that management can direct its attention to the difference from the planned amounts. If we assume that a company uses the perpetual inventory system and that it carries all of its inventory accounts at standard textile cost then the standard textile cost of a finished product is the sum of the standard textile costs of the inputs:

- Direct material
- Direct labour
- Manufacturing overhead
  - Variable manufacturing overhead
  - Fixed manufacturing overhead

VARIANCE ANALYSIS

Variance analysis is usually associated with explaining the difference (or variance) between actual textile costs and the standard textile costs allowed for the good output. For example, the difference in materials textile costs can be divided into a materials price variance and a materials usage variance.

The difference between the actual direct labour textile costs and the standard direct labour textile costs can be divided into a rate variance and an efficiency variance. The difference in manufacturing overhead can be divided into spending, efficiency, and volume variances. Mix and yield variances can also be calculated. Variance analysis helps management to understand the present textile costs and then to control future textile costs. Variance analysis is also used to explain the difference between the actual sales dollars and the budgeted sales dollars. Examples include sales price variance, sales quantity variance, and sales mix variance. A difference in the relative proportion of sales can account for some of the difference in a company's profits.

RESPONSIBILITY ACCOUNTING

Responsibility accounting is an underlying concept of accounting performance measurement systems. The basic idea is that large diversified organizations are difficult, if not impossible to manage as a single segment, thus they must be decentralized or separated into manageable parts. These parts, or segments are referred to as responsibility centers that include: 1) revenue centers, 2) textile cost centers, 3) profit centers and 4) investment centers. This approach allows responsibility to be assigned to the segment managers that have the greatest amount of influence over the key elements to be managed. These elements include revenue for a revenue center textile costs for a textile cost center a measure of profitability for a profit center and return on investment for an investment center.

Controllability Concept

An underlying concept of responsibility accounting is referred to as controllability. Conceptually, a manager should only be held responsible for those aspects of performance that he or she can control. In my view, this concept is rarely, if ever, applied successfully in practice because of the system variation present in all systems. Attempts to apply the controllability concept produce responsibility reports where each layer of management is held responsible for all subordinate management layers.

Advantages and Disadvantages

Responsibility accounting has been an accepted part of traditional accounting control systems for many years because it provides an organization with a number of advantages. Perhaps the most compelling argument for the responsibility accounting approach is that it provides a way to manage an organization that would otherwise be unmanageable. In addition, assigning responsibility to lower level managers allows higher level managers to pursue other activities such as long term planning and policy making.

It also provides a way to motivate lower level managers and workers. Managers and workers in an individualistic system tend to be motivated by measurements that emphasize their individual performances. However, this emphasis on the performance of individuals and individual segments creates what some critics refer to as the "stovepipe organization." Others have used the term "functional silos" to describe the same idea. Information flows vertically, rather than horizontally. Individuals in the various segments and functional areas are separated and tend to ignore the interdependencies within the organization. Segment managers and individual workers within segments tend to compete to optimize their own performance measurements rather than working together to optimize the performance of the system.

An implicit assumption of responsibility accounting is that separating a company into responsibility centers that are controlled in a top down manner is the way to optimize the system. However, this separation inevitably fails to consider many of the interdependencies within the organization. Ignoring the interdependencies prevents teamwork and creates the need for buffers such as additional inventory, workers, managers and capacity. Of course, a system that prevents teamwork and creates excess is inconsistent with the lean enterprise concepts of just-in-time and the theory of constraints. For this reason, critics of traditional accounting control systems advocate managing the system as a whole to eliminate the need for buffers and excess. They also argue that companies need to develop process oriented learning support systems, not financial results, fear oriented control systems.

The information system needs to reveal the company's problems and constraints in a timely manner and at a disaggregated level so that empowered users can identify how to correct problems, remove constraints and improve the process. According to these critics, accounting control information does not qualify in any of these categories because it is not timely, disaggregated, or user friendly. This harsh criticism of accounting control information leads us to a very important controversial question. Can a company successfully implement just-in-time and other continuous improvement concepts while retaining a traditional responsibility accounting control system?

Although the jury is still out on this question, a number of field research studies indicate that accounting based controls are playing a decreasing role in companies that adopt the lean enterprise concepts. In a recent study involving nine companies, each company answered this controversial question in a different way by using a different mix of process oriented versus results oriented learning and control information. Since each company is different, a generalized answer to this question for all firms in all situations cannot be provided.

3

Functional Suitability of Soluble Peroxidases from Easily control in Decolorization of Synthetic Dyes

With the growing commercial availability of dyes, their range and scope of application is also expanding. Consequently, a large amount of unused dyes are released in the industrial effluents. Ninety per cent of textile dyes entering modern activated sludge sewage treatment plants pass through unchanged. There is an intense environmental concern about the fate of these unbound dyes. These discharged dyes form toxic products and their strong colour causes turbidity which even at very low concentrations has a huge impact on the aquatic environment. These synthetic reactive dyes bond covalently with fabric and contain chromophoric groups like anthraquinone, azo, triarylmethane etc. along with reactive groups *viz.*, vinyl sulphone, chlorotriazine, trichloropyrimidine etc. (Sumathi and Manju, 2000; Keharia and Madamvar, 2003). Disperse dyes and acid dyes have low solubility in water. They are mainly used in the dyeing of polyesters and find minor use in dyeing cellulose acetates and polyamides. Azo dyes constitute the largest group of colorants used in industry. These dyes can be precipitated or adsorbed only in small amounts, while under anaerobic conditions they are cleaved by microorganisms forming potentially carcinogenic aromatic amines (Chung and Cerniglia, 1992). Several studies on decolorization of textile dyes used in industries have been conducted. Some non-textile dyes commonly present in industrial effluents have been studied. Dye removal from wastewaters with traditional physicochemical processes, such as coagulation, adsorption and oxidation with ozone is expensive, can generate large volumes of sludge and usually requires the addition of environmentally hazardous chemical additives (Chen, 2006). On the other hand, most of the synthetic dyes are xenobiotic compounds which are poorly removed by the use of conventional biological aerobic treatments (Marco et al., 2007a, b). Although, biodegradation appears to be a promising technology, unfortunately the analysis of contaminated soil and water has shown persistence of toxic pollutants even in the presence of microorganisms (Robinson et al., 2001; Keharia and Madamvar, 2003).

Decolorization of dye wastewater is an area where innovative treatment technologies need to be investigated. The focus in recent times has shifted towards enzyme based treatment of colored wastewater/industrial textile effluents. The peroxidase and polyphenol oxidases participate in the degradation of a broad range of substrate even at very low concentration. Further, these peroxidases and polyphenol oxidases have been used for treatment of dyes but large scale exploitation has not been achieved due to their low enzymatic activity in biological materials and high cost of purification (Bhunia et al., 2001; Shaffiqu et al., 2002; Verma and Madamwar, 2002). Bioremediation is a viable tool for restoration of contaminated subsurface environments. It is gaining importance due to its cost effectiveness, environmental friendliness and production of less sludge as compared to chemical and physical decomposition processes. Here too microbial treatment has certain inherent limitations (Husain and Jan, 2000; Duran and Esposito, 2000; Torres et al., 2003). It has been shown that peroxidases catalyze a variety of oxidation reactions and importantly dyes recalcitrant to peroxidase shows significant decolorization in the presence of Redox mediators (Calcaterra et al., 2008). Redox mediated enzyme catalysis has wide application in degradation of polycyclic aromatic hydrocarbons which includes phenols, biphenyls, pesticides, insecticides etc. (Husain and Husain, 2008; Calcaterra et al., 2008).

PEROXIDASES IN DYE DECOLORIZATION WITH REDOX MEDIATORS

Enzymatic approach has gained considerable interest in the decolorization/ degradation of textile and other industrially important dyes present in wastewater. This strategy is ecofriendly and useful in comparison to conventional chemical, physical and biological treatments, which have inherent serious limitations. Enzymatic treatment is very useful due to the action of enzymes on pollutants even when they are present in very dilute solutions and recalcitrant to the action of various microbes participating in the degradation of dyes. Several enzymes (peroxidases, manganese peroxidases, lignin peroxidases, laccases, microperoxidase-11, polyphenol oxidases and azoreductases) have been evaluated for their potential in decolorization and degradation of dyes. Although, some recalcitrant dyes are not degraded/ decolorized in the presence of such enzymes, the addition of certain Redox mediator enhances the range of substrates and efficiency of degradation of the recalcitrant compounds. However, very few Redox mediators are frequently used which includes 1- hydroxybenzotriazole (HOBT), veratryl alcohol, violuric acid (VA), 2-methoxyphenothiazone. The enzymes in soluble form cannot be exploited on large scale due to limitations of stability and reusability and consequently, the use of immobilized enzymes has significant advantages over soluble enzymes. In the near future, technology based on the enzymatic treatment of dyes present in the industrial effluents/wastewater will play a vital role. Treatment of wastewater on a large scale will also be possible by using reactors containing immobilized enzymes.

DYE DECOLORIZATION WITH TURNIP (*BRASSICA RAPA*) PROTEINS

The potential of partially purified turnip (*Brassica rapa*) proteins have been studied by Matto and Husain, (2007) on decolorization of certain direct dyes like Direct Red 23, Direct Red 239, Direct Blue 80 and Direct Yellow 4. The turnip proteins showed enhanced decolorization in the presence of Redox mediators. Redox mediators explored for dye decolorization of these dyes were HOBT, alpha naphthol, vanillin, L-histidine, VA, catechol, quinol, bromophenol, 4-nitrophenol and gallic acid. Six out of 10 investigated compounds showed their potential in enhancing the decolorization of direct dyes. The performance of each was evaluated at different concentrations of mediator and enzyme. The decolorization of all tested direct dyes was maximum in the presence of 0.6 mM Redox mediator at pH 5.5and 30°C. Complex mixtures of dyes decolorized maximally in the presence of 0.6 mM Redox mediator (HOBT/VA). In order to examine the operational stability of the enzyme preparation, the enzyme was exploited for the decolorization of mixtures of dyes for different times in a stirred batch process. There was no change in decolorization of an individual dye or their mixtures after 60 min of incubation; the enzyme caused more than 80 per cent decolorization of all dyes in the presence of 1-hydroxybenzotriazole/violuric acid. However, there was no desirable increase in dye decolorization of the mixtures on overnight incubation. The treatment of such polluted water in the presence of Redox mediators caused the formation of insoluble precipitate, which could be removed by the process of centrifugation. Such catalyzed oxidative coupling reactions may be important for natural transformation pathways for dyes and indicate their potential use as an efficient means for removal of dyes colour from waters and wastewaters.

DYE DECOLORIZATION WITH FENUGREEK (*TRIGONELLA FOENUM-GRAECUM*) SEED PROTEINS

Peroxidase from fenugreek (*Trigonella foenum-graecum*) seeds is highly effective in the decolorization of textile effluent. The role of six Redox mediators has been investigated for effective decolorization by FSP (Fenugreek Seed Proteins). The maximum decolorization of textile effluent was observed in the presence of 1.0 mM 1-hydroxybenzotrizole, 0.7 mM H_2O_2, and 0.4 U/ml of FSP in the buffer of pH 5.0 at 40°C in 2.5 h. The decolorization of textile effluent in a batch process by peroxidase is 85 per cent in 5 h, whereas the complete decolorization of textile effluent by membrane entrapped FSP was observed within 11 h of its operation. The absorption spectra of treated effluent exhibited a marked diminution in the absorbance at different wavelengths compared to untreated effluent. The removal of colored aromatic compounds from wastewater by peroxidases is well-known. These enzymes are currently being employed for the treatment of aromatic compounds (Husain and Husain, 2008; Husain et al., 2009). Most of the studies on dye decolorization have been done in a defined media or synthetic wastewater where a single dye or their mixtures are usually present. However, industrial effluents are more complex due to the presence of other contaminating substances along with colored compounds; under such conditions, the treatment of these pollutants is a difficult problem (Husain et al., 2010). Textile effluents alone are recalcitrant to the action of FSP, but in the presence of Redox mediators, they decolorized significantly. Many other aromatic pollutants such as aromatic amines, dyes and bisphenol A have already been oxidized by peroxidases in the presence of Redox mediators (Karim and Husain, 2009; Matto and Husain, 2008). The oxidation of effluent in the presence 1.0 mM HOBT is quite low compared to the Redox mediators used in earlier studies to treat various colored pollutants (Rodriguez-Couto et al., 2005; Matto and Husain, 2008). The high concentrations of Redox mediators may not be appropriate for wastewater treatment because of high cost of mediators or possibility of creating negative impacts on effluent toxicity or in the environment upon their disposal into receiving waters (Kurniawati and Nicell, 2007). The maximum decolorization of textile effluent with FSP was at 0.7 mM concentration of H_2O_2. The concentration of H_2O_2 >1.2 mM acted as an inhibitor of peroxidase activity possibly by causing irreversible oxidation of enzyme ferri-heme group which is essential for its activity (Vazquez-Duarte et al., 2001). The differences in time course of removal of various dyes might be due to the structural barrier and the electron localization among them (Jauregui et al., 2003). The maximum decolorization (68 per cent) of textile effluent was obtained by 0.4 U/ml of FSP. FSP catalyzed the decolorization of effluent over a wide range of pH with an optimum at pH 5.0. The maximum decolorization of effluent was at a temperature of 40°C. FSP successfully removed more than 85 per cent colour of the effluent in 5 h. After 5 h, the increase in the rate of effluent decolorization has not been recorded. Complete removal of colour has been studied by using enzyme in a membrane bag. This type of reactor would increase the efficiency of dye decolorization because enzyme can be reused as it did not mix with the rest of the treated solution. Some workers reported that *Pseudomonas putida* CCRC14365 peroxidase was capable of removing phenol in hollow-fibre membrane bioreactor. *P. putida* fully degraded 2,000 mg/L of phenol within 73 h; even at a level of 2,800 mg/L, phenol could be degraded by more than 90 per cent after 95 h of operation (Chung et al., 2004). In another report, a novel tube membrane bioreactor has been used for the treatment of an industrially produced wastewater arising in the manufacture of 3-chloronitrobenzene and nitrobenzene; in 1.7 h, over 99 per cent of each 3-chloronitrobenzene and nitrobenzene from the wastewater was degraded

(Livingston, 2004). For the decolorization and removal of colored compounds from textile effluent, spectral analysis is an important aspect to demonstrate a loss in these compounds after treatment with enzyme. The decrease in absorbance peaks in the UV-visible region is a strong evidence for the decolorization of aromatic pollutants from wastewater. Some earlier spectral analyses of the enzymatically treated aromatic amines and their mixtures have shown a significant decrease in absorbance (Kurniawati and Nicell, 2007). There was 80 per cent decolorization of the chromophoric groups and a significant reduction in the peak associated with the aromatic ring. The UV-visible absorbance spectrum for diluted textile effluent has been taken before and after treatment by FSP in the presence of various Redox mediators. The treatment of effluent by FSP in the presence of Redox mediator, HOBT, produces insoluble aggregates which can successfully be removed by centrifugation. The peroxidase from fenugreek seeds has potential in the remediation of hazardous aromatic pollutant and its use could be extended to the large-scale treatment of textile effluents and other related aromatic compounds by employing more effective and cheaper Redox mediators.

DYE DECOLORIZATION WITH HORSE RADISH PEROXIDASE (HRP)

HRP is extracted from horse radish roots, and its performance has been evaluated in soluble and immobilized form by conducting batch experiments in the presence of H_2O_2 (Vasantha et al., 2006). The oxidation of Direct Yellow 12 dye has been tested as a function of HRP at fixed concentration of H_2O_2, and at constant HRP activity (1.8 units/ml). The optimum contact time required for dye removal was 1 h 45 min for vials containing 5 ml of dye solution (10 mg/L), 1.8 units of enzyme 1.5 ml/L of H_2O_2 were added and the reaction mixture (24°C, pH 4) was agitated for 2 h 15 min. After this period, dye removal was negligible up to remaining 1 h 35 min. The Direct Yellow 12 dye at varying aqueous-phase pH of the reaction mixture between 2 and 10 by keeping the dye concentration at 10 mg/L, enzyme concentration at 1.8 units, H_2O_2 dose at 1.5 ml/L, reaction temperature at 24°C and the contact time (1 h 45 min) constant exhibited 70 per cent of the dye removal at an aqueous-phase pH of 4. The dye removal dropped significantly from pH 5 to 8 and the same trend continued up to an aqueous phase of pH 10. Aqueous phase of pH 4 resulted in higher HRP activity compared to other pH ranges from 3 to 9. Hydrogen peroxide acts as a co-substrate to activate the enzymatic action of peroxide radical. It contributed in the catalytic cycle of peroxidase oxidizing the native enzyme to form an enzymatic intermediate which accepts the aromatic compound to carry out its oxidation to form a free radical form. Experiments were carried out to find out the optimum H_2O_2 concentration required to bring out the conversion of dye by varying the H_2O_2 dose from 1 to 3 ml/L in the reaction mixture by keeping all the other experimental conditions constant (dye concentration 10 mg/L; temperature 24°C; enzyme concentration 1.8 units; reaction time 1 h 45 min). Their findings indicate maximum dye removal at H_2O_2 concentration of 2 ml/L and similar degradation with 2.5 and 3 ml/L; therefore, 2 ml/L was taken as the optimum H_2O_2 dose for dye removal (Vasantha et al., 2006). The concentration of the substrate present in the aqueous phase significantly influences enzyme-mediated reaction. If the amount of enzyme concentration is kept constant and the substrate concentration is gradually increased the reaction will increase until it reaches maximum. After obtaining the equilibrium state, further addition of the substrate will not change the rate of reaction. Studies carried out at different concentrations of the dye, *i.e.* 5–40 mg/L, keeping the other parameters constant indicated a dye concentration of 25 mg/L to be the cut-off concentration of the dye for optimum removal at the specified experimental conditions along with retaining the activity of HRP enzyme and to protect the protein from denaturation. Application of free enzyme in industrial process is not economically viable, while immobilization/ entrapment of enzyme results in repeated application. Two types of polymeric materials—alginate and acryl amide—have been used to study their relative efficiency in dye removal for the entrapment of peroxidase. Normally, enzyme immobilization is expected to provide stabilization effect restricting the protein unfolding process as a result of the introduction of random intra- and intermolecular crosslinks. Zille et al. (2003) have reported less availability of the enzyme for interaction with anionic dyes due to the immobilization in a particular matrix. For the immobilization of HRP enzyme acryl amide gel was more efficient in dye removal compared to alginate matrix. About 78 per cent of dye removal was observed with acryl amide gel-immobilized beads, while with alginate matrix it was only 52 per cent. Gel-immobilized HRP was effective in dye removal compared to free HRP (69 per cent), and alginate-immobilized HRP showed inferior performance. The effect of aqueous-phase pH on the enzyme-catalyzed degradation with alginate- and acryl amide-immobilized HRP enzyme suggests that after pH 4 there is a decrease in the dye removal capacity for both types of the entrapped matrices. About 78 per cent and 54 per cent of the dye removal was observed at an aqueous phase of pH 4 for acryl amide and alginate-entrapped beads respectively. The relative inferior performance of alginate-immobilized HRP compared to acryl amide may perhaps be due to lesser availability of the peroxidase structure to the dye molecule in the alginate mix compared to acryl amide. The effective performance of acryl amide-entrapped beads may be attributed to the non-ionic nature of the beads, which results in minimum modification of the enzyme properties and unaffected nature of the charged substrate as well as product diffusion. The electron-withdrawing nature of the azo linkages obstructs the susceptibility of azo dye molecules to oxidative reactions. Only specialized azo dye-reducing enzymes could degrade azo dyes. In addition, when Direct Yellow

12 dye was reduced using the enzymatic method, the oxidation capacity increased with increasing concentrations of HRP and H_2O_2 at pH 4. Gel/alginate based enzyme immobilization reduced the azo dye. The immobilized enzyme beads could further be used two–three times for the removal of the same dye with lower efficiency. The application of enzyme-based systems in waste treatment is unusual, given that many drawbacks are derived from their use, including low efficiency, high costs and easy deactivation of the enzyme.

DYE COLOUR REMOVAL WITH BITTER GOURD (*MOMORDICA CHARANTIA*) PEROXIDASE

Application of partially purified BGP (Bitter Gourd Peroxidase) in the decolorization of textile and other industrially important dyes has been explored. Simple ammonium sulphate precipitated proteins from bitter gourd were taken for the treatment of a number of dyes present in polluted wastewater (Akhtar et al., 2005a, b). Partially purified preparation of BGP was obtained by adding 20–80 per cent ammonium sulphate and this preparation exhibited a specific activity of 99.0 EU (Enzyme units) of peroxidase/mg protein. Peroxidases from bitter gourd were highly stable against pH, heat, urea, water miscible organic solvents, detergents and proteolysis. The dye solution was stable upon exposure to H_2O_2 or to the enzyme alone. The dye precipitation was due to H_2O_2-dependent enzymatic reaction, possibly involving free-radical formation followed by polymerization and precipitation. Eight textile reactive dyes were tested with increasing concentration (0.133–0.339 EU of BGP/ml) of reaction volume for 2 h at 37°C. The decolorization of Reactive Red 120, Reactive Blue 4, Reactive Blue 160 and Reactive Blue 171 was continuously enhanced by adding increasing concentration of BGP. Reactive Blue 4 was completely decolorized with 0.399 EU of BGP/ml of reaction mixture in the absence of HOBT. Reactive Orange 4, Reactive Orange 86, Reactive Red 11 and Reactive Yellow 84 were recalcitrant to the BGP action. Four reactive textile dyes were treated by step-wise addition of enzyme, adding 0.133 EU of BGP/ml of reaction volume at each step after every 1 h, to the decolorizing solution. Each enzyme addition exhibited rapid disappearance of colour. Reactive Blue 160 was completely decolorized after the third enzyme addition while Reactive Blue 4 was almost decolorized completely after the second enzyme addition. Reactive Blue 171 and Reactive Red 120 were decolorized to 56 per cent and 39 per cent, respectively after the third enzyme addition. However, Reactive Red 11, Reactive Orange 4, Reactive Orange 86 and Reactive Yellow 84 were recalcitrant to decolorization by BGP. Of the reactive dyes incubated with 0.266 EU of BGP/ml of reaction volume for increasing time period only four dyes decolorized on treatment with BGP for 1h at 37°C. Although more colour disappeared on incubation for longer duration, the rate of decolorization was slow. All the four dyes were decolorized with varying per centages (30–90 per cent). The decolorization of dyes with 0.266 EU of BGP/ml of reaction volume on incubation at 37°C for 4 h was 88 per cent for Reactive Blue 4, 71 per cent for Reactive Blue 160, 31 per cent for Reactive Blue 171 and 28 per cent for Reactive Red 120. Rest of the four dyes: Reactive Red 11, Reactive Orange 4, Reactive Orange 86 and Reactive Yellow 84 were fully recalcitrant to decolorization by BGP even after 4 h incubation with similar treatment. Thirteen different non-textile dyes were treated with 0.167 EU of BGP/ml of reaction volume at 37°C for 1h. Carmine, Methyl Orange, Methylene Blue, Coomassie Brilliant Blue G-250, Rhodamine 6G, Methyl Violet 6B, 1:2 naphthaquinone 4-sulphonic acid and Martius Yellow dyes were recalcitrant to decolorization by the BGP action or were slowly decolorized during the progress of the reaction. Maximum decolorization achieved by partially purified BGP was 53 per cent for Naphthalene Black 12B, 57 per cent for Coomassie Brilliant Blue R 250, 96 per cent for Evans Blue, 51 per cent for Eriochrome Black T and 86 per cent for Celestine Blue. Twenty-one dyes used in this study were treated with BGP in the presence of 1.0 mM HOBT and 0.6 mM H_2O_2 at 37°C. Presence of HOBT drastically enhanced the rate of decolorization of recalcitrant dyes. Reactive Orange 4, Reactive Red 11, Reactive Yellow 84 and Reactive Orange 86 were recalcitrant to decolorization in the absence of HOBT. However, these dyes were decolorized up to 98 per cent, 80 per cent, 70 per cent and 78 per cent, respectively by the action of 0.399 EU/ml of BGP in the presence of 1.0 mM HOBT at 37°C for 2 h. Non-textile dyes *e.g.*, Carmine, Methyl Orange, Coomassie Brilliant Blue G250, Rhodamine 6G, Methylene Blue and Methyl Violet 6B, which were recalcitrant to decolorization by BGP in absence of HOBT, were almost completely decolorized in presence of 1.0 mM HOBT. However, 1:2-naphthaquinone 4-sulphonic acid was recalcitrant to decolorization even in the presence of HOBT. Decolorization of textile reactive dyes in the absence of HOBT occurred by the formation of precipitate, which settled down and removed by centrifugation. Several earlier investigators have shown that the treatment of phenols and aromatic amines by peroxidases and tyrosinases resulted in the formation of large insoluble aggregates. However, the decolorization of textile and other dyes by BGP in presence of 1.0 mM HOBT appeared without the formation of any precipitate. It suggested that the decolorization of dyes took place via degradation of aromatic ring of the compounds or by cleaving certain functional groups as reported elsewhere (Christian et al., 2003; Husain et al., 2009). HOBT could have a dual role, first as a mediator by increasing the substrate range of dyes for BGP and second enhancing the rate of oxidation. Complex mixtures of various reactive textile and other industrially important dyes have been studied by mixing three or four different dyes in equal proportions and incubated with 0.266 EU of BGP/ ml in the presence of 1.0 mM HOBT and 0.6 mM H_2O_2 for 1h at 37°C. Wave length maxima for each dye mixture were determined and decolorization of mixtures was monitored after the incubation period. All the mixtures decolorized by more than 80 per cent. The decolorization rate of mixtures of dyes was slower than that of pure dye solution. This supports an earlier observation that the biodegradation of various phenols in the form of mixtures was quite slow compared to the independent phenol (Kahru et al., 2000). Bourbonnais and Paice (1990) described for the first time the use of Redox mediators by allowing laccase to oxidize non-phenolic compounds thereby expanding the range of substrates that can be oxidized by this enzyme. The mechanism of action of laccase mediator system has been extensively studied and it is used in the textile industry in the finishing process for indigo stained materials. Several workers have demonstrated that the use of Redox mediator system enhanced the rate of dye decolorization by several folds but these mediators were required in very high concentrations (5.7 mM violuric acid/ laccase system, 11.6 mM of HOBT/laccase system) (Soares et al., 2001a,b; Claus et al., 2002). For the first time, it was shown that the decolorization of dyes by BGP was effective at very low concentration of HOBT (1.0 mM). The rate of non-textile dyes decolorization was also enhanced by 2–100 folds (Akhtar et al., 2005a). Several reports indicate the enhancement of laccase activity by free radical mediators; however the enhancement of peroxidase activity by a Redox mediator has been studied for the first time. Peroxidase/ Redox mediator system will prove a sensitive and an inexpensive procedure for the treatment of dyes present in complex mixtures/industrial effluent.

DYE COLOUR REMOVAL WITH TOMATO (*LYCOPERSICON ESCULENTUM*) PEROXIDASE

Matto and Husain (2009) investigated the role of concanavalin A (Con A)-cellulose-bound tomato peroxidase for the decolorization of direct dyes. Cellulose was used as an inexpensive material for the preparation of bioaffinity support. Con A-cellulose-bound tomato peroxidase exhibited higher efficiency in terms of dye decolorization as compared to soluble enzyme under various experimental conditions. Both Direct Red 23 and Direct Blue 80 dyes were recalcitrant to the action of enzyme without a Redox mediator. Six compounds were investigated for Redox-mediating property. Immobilized peroxidase decolorized both dyes to different extent in the presence of all the used Redox mediators. However, 1- hydroxybenzotriazole emerged as a potential Redox mediator for tomato peroxidase catalyzed decolorization of direct dyes. These dyes were maximally decolorized at pH 6.0 and 40°C by soluble and immobilized peroxidase. The absorption spectra of the untreated and treated dyes exhibited a marked difference in the absorption at various wavelengths. Immobilized tomato peroxidase showed a lower Michaelis constant than the free enzyme for both dyes. Soluble and immobilized tomato peroxidase exhibited significantly higher affinity for Direct Red 23 compared to Direct Blue 80.

DYE DECOLORIZATION WITH POINTED GOURD (*TRICHOSANTHES DIOICA*) PEROXIDASE

The effects of *Trichosanthes dioica* peroxidase along with Redox mediators on decolorization of water insoluble disperse dyes; Disperse Red 19 (DR19) and Disperse Black 9 (DB9) have been studied by Jamal et al., (2010). Nine different Redox mediators; bromophenol, 2, 4-dichlorophenol, guaiacol, 1-hydroxybenzotriazole, m-cresol, quinol, syringaldehyde, vanillin and violuric acid were evaluated. Among the chosen mediators, 1-hydroxybenzotriazole was most effective for decolorization with PGP (Pointed Gourd Peroxidase). At a concentration of 0.45U/mL the peroxidase could decolorize Disperse Red 19 to a maximum of 79 per cent with 0.2mM 1-hydroxybenzotriazole whereas Disperse Black 9 decolorized upto 60 per cent with 0.5mM 1-hydroxybenzotriazole. The time, pH and temperature at which maximum decolorization were recorded was 60 min, 4 and 42°C. It has been observed that the dye solutions were recalcitrant upon exposure to HOBT, H_2O_2 or to the enzyme alone but the enzyme in the presence of Redox mediators was much effective in performing the decolorization of the dyes, implying dye decolorization was a result of Redox mediated H_2O_2-dependent enzymatic reaction. It has already been reported that Redox mediators have the potential to mediate an oxidation reaction between a substrate and an enzyme.

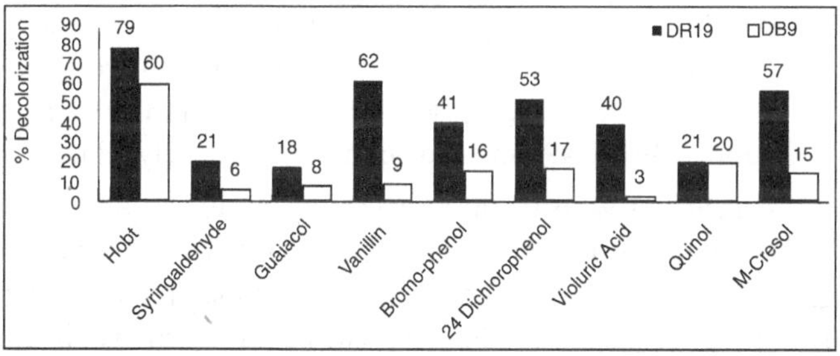

Fig. Per cent Dye decolorization as a function of different Redox mediators. The dyes DR19 (25mgL^{-1},5.0mL) and DB9(50 mgL^{-1},5.0mL) solutions were incubated independently with PGP (0.45 Uml^{-1}) in the presence of 0.5mM concentration of each Redox mediators; other conditions were 0.8mM H_2O_2, 100mM glycine HCl buffer, pH 4.0 for 60 min at 37°C. (λmax for DR19 and DB9 are 495nm and 461nm respectively) [Jamal et al., 2010]

Different Redox mediators have different mediation efficiency which is governed by Redox potential of the mediator and the oxidation mechanism of the substrate. Oxidation of substrate occurs by free radical formation by the mediator. The free radicals can be formed either by one-electron oxidation of substrate or by abstraction of a proton from the substrate. The pointed gourd peroxidase was effective in decolorizing the dyes at low concentrations of HOBT. Although the extent of decolorization of DR19 and DB9 increased with increasing concentrations of HOBT, the maximum decolorization was observed to be 79 per cent and 60 per cent with 0.2 mM and 0.5 mM for DR19 and DB9, respectively. Further addition of HOBT resulted in a slow decrease in decolorization of both the dyes. This inhibition could likely be due to the high reactivity of HOBT radical, which might undergo chemical reactions with side chains of aromatic amino acid by enzyme thereby; inactivating it. Hence, the dosage of Redox mediator is an important factor for the enzyme-mediated decolorization.

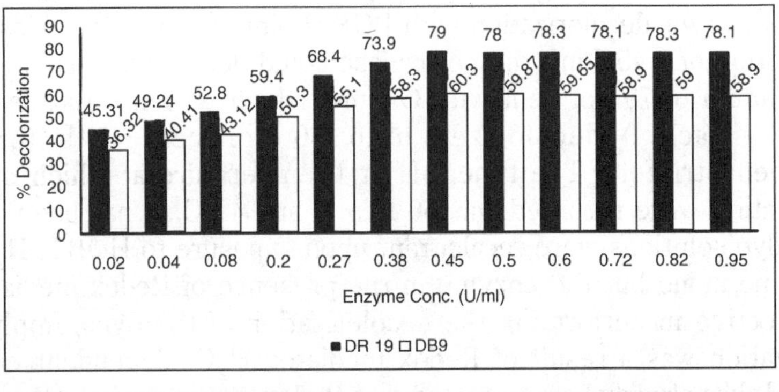

Fig. Per cent Dye decolorization as a function of different enzyme (PGP) concentrations. The dyes DR19 (25mgL^{-1},5.0mL) and DB9(50 mgL^{-1},5.0mL)solutions were incubated independently with PGP (0.02 to 0.95 Uml^{-1}) in the presence of 0.2mM and 0.5mMconcentration of HOBT for DR19 and DB9 respectively; other conditions were 0.8mM H_2O_2, 100mM glycine HCl buffer, pH 4.0 for 60 min at 37°C. (λmax for DR19 and DB9 are 495nm and 461nm respectively) [Jamal et al., 2010].

The enzyme reacted well to decolorize both the dyes in the presence of 0.8 mM H_2O_2. The maximum decolorization was obtained at 0.8mM H_2O_2 which is slightly higher than reported for soybean peroxidase, bitter gourd peroxidase (BGP) and turnip peroxidase. The reaction temperature is an important parameter which effects the decolorization of dyes. The maximum decolorization for both the dyes DR19 and DB9 was at 42°C. It has been reported that BGP mediated disperse dye decolorization was optimal at 40°C.

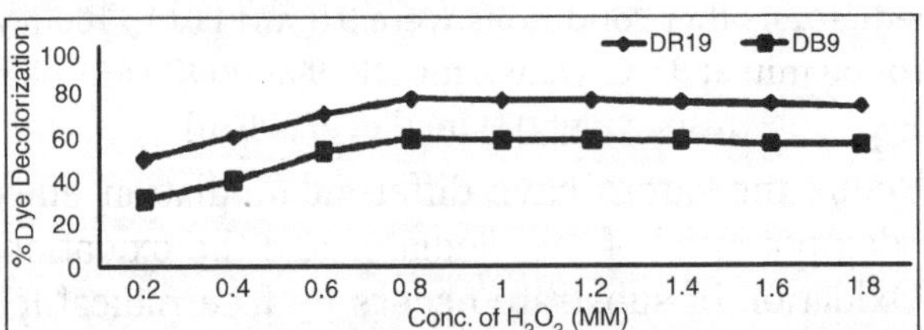

Fig. Per cent Dye decolorization at different concentrations of H_2O_2. The dyes DR19 (25mgL^{-1},5.0mL) and DB9(50 mgL^{-1},5.0mL) solutions were incubated independently with H_2O_2 concentrations and PGP (0.45 Uml^{-1}) in the presence of varying amounts of H_2O_2 (0.2 to 1.8 mM) for DR19 and DB9 respectively; and other conditions were 100mM glycine HCl buffer, pH 4.0 for 60 min at 37°C. (λmax for DR19 and DB9 are 495nm and 461nm respectively) [Jamal et al., 2010].

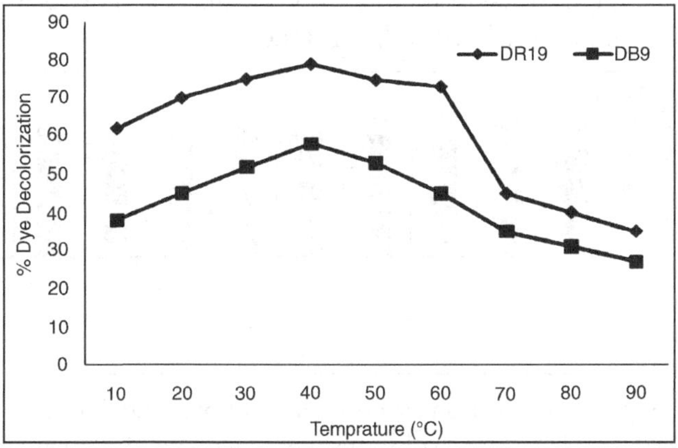

Fig. Per cent Dye decolorization as a function of temperature. The dyes DR19 (25mgL^{-1},5.0mL) and DB9 (50 mgL^{-1},5.0mL)solutions were incubated independently with PGP (0.45 Uml^{-1}) in the presence of HOBT, 0.8mM H_2O_2, 100mM glycine HCl buffer, pH 4.0 for 60 min at temperatures (20°C to 90°C). (λmax for DR19 and DB9 are 495nm and 461nm respectively) [Jamal et al., 2010].

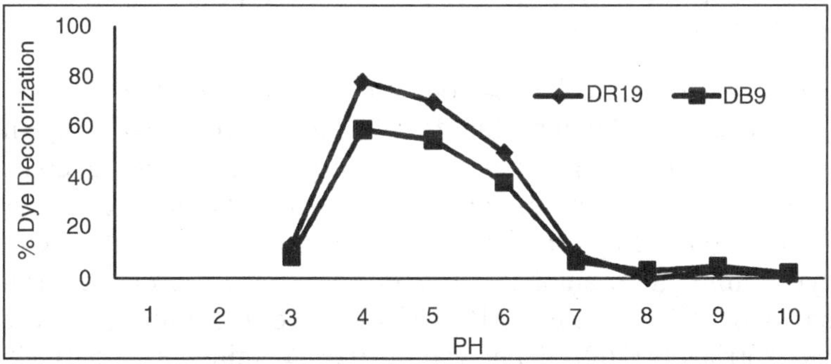

Fig. Per cent Dye decolorization as a function of pH. The dyes DR19 (25mgL^{-1}, 5.0mL) and DB9 (50 mgL^{-1},5.0mL) solutions were incubated independently with PGP (0.45 Uml^{-1}) in the presence of HOBT, 0.8mM H_2O_2, 100mM glycine HCl buffer at different pH (2,3,4,5,6,7,8,9 and 10) for 60 min at 37°C. (λmax for DR19 and DB9 are 495nm and 461nm respectively) [Jamal et al., 2010].

The maximum decolorization of DR19 and DB9 was obtained at an acidic of pH 4.0. It has earlier been reported that the degradation of industrially important dyes by enzymes such as horse radish peroxidase, polyphenol oxidase, BGP and laccase was also maximum in the buffers of acidic pH. DR19 and DB9 were maximally decolorized within 20 min of incubation. There was slow and gradual enhancement of decolorization upto 60 min of incubation. It was also evident from the observation that DR19 was decolorized to a greater extent within 20 min in the presence of only 0.2 mM HOBT. However, DB9 was decolorized maximally in the presence of 0.5 mM HOBT and decolorization rate was slow. This is consistent with reports that decolorization rate varies, depending upon the type of dye to be treated.

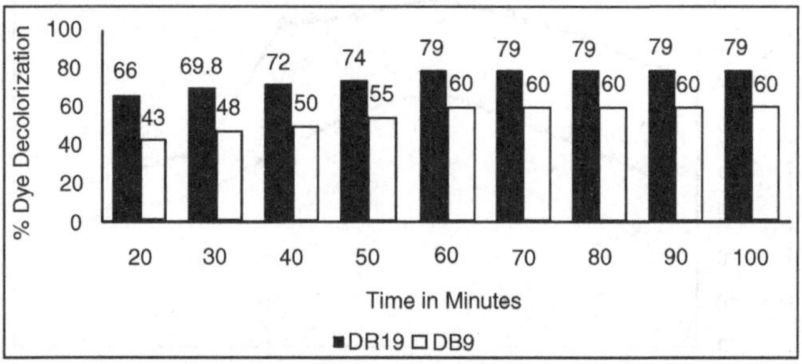

Fig. Per cent Dye decolorization as a function of time. The dyes DR19 (25mgL^{-1}, 5.0mL) and DB9(50 mgL^{-1},5.0mL) solutions were incubated independently with PGP (0.45 Uml^{-1}) in the presence of HOBT, 0.8mM H_2O_2, 100mM glycine HCl buffer, pH 4.0 and for time 20 to 100 min at 37°C. (λmax for DR19 and DB9 are 495nm and 461nm respectively) [Jamal et al., 2010].

TRICHOSANTHES DIOICA PROTEINS IN DECOLORIZING INDUSTRIALLY IMPORTANT TEXTILE, NON-TEXTILE DYES AND DYE MIXTURES

In one of my recent works, the competency of *Trichosanthes dioica* proteins in decolorizing industrially important textile, non-textile dyes and dye mixtures in the presence of Redox mediators under varying experimental conditions of pH, temperature, time intervals and enzyme concentration on the basis of one-factor-at-a-time (OFAT) method has been studied. All the textile dyes and non-textile dyes, ammonium sulphate, and Tween-20 were procured from Sigma Chemical Co. (St. Louis, MO, USA). Redox mediator's *viz.*, 1-hydroxybenzotriazole (HOBT) and vanillin were obtained from SRL Chemicals (Mumbai, India). All other chemicals were of analytical grade. The pointed gourds were purchased from the local market. The samples were aseptically transferred into sterilized plastic bags. Briefly, 100 g of pointed gourd was homogenized in 180 ml of 100 mM sodium acetate buffer, pH 5.6. The homogenate was filtered through multi-layers of cheese cloth and then centrifuged at the speed of 10,000 × g on a Remi C-24 cooling centrifuge for 30 min at 4°C. The clear solution thus obtained was used for salt fractionation by adding 10 per cent to 80 per cent (w/v) $(NH_4)_2SO_4$. The proteins were precipitated by continuously stirring at 4°C overnight. The precipitate was collected by centrifugation at 10,000 × g on a Remi C-24 cooling centrifuge, dissolved in 100 mM sodium acetate buffer, pH 5.6 and dialyzed against the assay buffer (0.1 M glycine HCl buffer, pH 4.0) (Akhtar et al., 2005a). Protein concentration was estimated by taking BSA as a standard protein and following the procedure of Lowry et al (1951). Peroxidase activity was determined by a change in the optical density (A_{460} nm) at 37°C by measuring the initial rate of oxidation of 6.0 mM o-dianisidine HCl in the presence of 18.0 mM H_2O_2 in 0.1 M glycine-HCl buffer, pH 4.0, for 20 min at 37°C. One unit of activity was defined as the amount of enzyme that transformed 1µmol of o-dianisidine HCl as substrate per min. The dyes (45-210 mg/l) were solubilized in 100 mM glycine HCl buffer, pH 4.0. Each dye was independently incubated with pointed gourd peroxidase (PGP) (0.45 EU/mL) in 100 mM glycine HCl buffer, pH 4.0 in the presence of 0.80 mM H_2O_2 for varying times at 37°C. The reaction was stopped by boiling at 100°C for 7 min. Dye decolorization was monitored by measuring the difference at the maximum absorbance for each dye as compared with control experiments without enzyme on UV-visible spectrophotometer (JASCO V-550, Japan). Untreated dye solution (inclusive of all reagents except the enzymes) was used as control (100 per cent) for the calculation of per cent decolorization. The dye decolorization was calculated as the ratio of the difference of absorbance of treated and untreated dye to that of treated dye and converted in terms of percentage.

Effect of Redox Mediators on Decolorization Profile of Textile and Non-textile Dyes

The effect of different Redox mediators on the dye decolorization by PGP is shown in Table. Out of the two different Redox mediators studied for dye decolorization, HOBT was more effective in decolorizing the dyes under study. The extent of decolorization in the presence of HOBT was in the range of 98.6 per cent to 69.8 per cent for the reactive dyes whereas the disperse dyes studied exhibited decolorization in the range of 79.2 per cent to 61.2 per cent. The effective HOBT concentrations were 1.0 mM and 0.2 mM for reactive and disperse dyes, respectively.

The dye decolorization with vanillin was 71.2 per cent to 60.2 per cent for reactive dyes and 55.3 per cent to 34.5 per cent for disperse dyes at 1.0 mM concentration. Table-1 shows the effect of increasing concentrations of HOBT and vanillin (0.05 to 1.5 mM) on textile and other dyes. With increasing concentration of HOBT or vanillin there was an increase in the extent of decolorization of both dyes.

However, the decolorization of each textile dye was lower at each of the varying concentration of vanillin than HOBT. The non-textile dyes were effectively decolorized at 1.0 mM of HOBT and vanillin independently but here too, influence of HOBT was substantial.

There was not much effect in per cent decolorization of the dyes above these concentrations of HOBT or vanillin.

Enzyme Activity Profile of *T. dioica* Peroxidase Mediated Decolorization of Textile Dyes

Figure-7a shows the extent of decolorization of reactive and disperse dyes with increasing concentration of PGP. The maximal decolorization for these two different classes of dyes was observed at PGP concentration of 0.45 EU/ml after an incubation time of 4 h. Dye decolorization was not significantly exhibited with any further increase of PGP.

The presence of 0.2 mM and 1.0 mM HOBT for decolorization of disperse and reactive dyes was effective in disappearance of Reactive Blue 15 and Reactive Orange 15 to the extent of 96.2 per cent and 94.6 per cent respectively whereas the others, Reactive Red 4 and Reactive Yellow 2 disappeared up to 88.2 per cent and 89.8 per cent respectively.

Among the disperse dyes only Disperse Black 9 was effectively decolorized up to 79 per cent whereas the others showed disappearance of colour below 69.3 per cent. In the presence of 1.0 mM vanillin, partially purified *T. dioica* peroxidase catalyzed decolorization at 0.45 EU/ml after an incubation period of 4 h in the range of 61.4 per cent to 71.3 per cent for the dyes under study. The decolorization profile for all the dyes significantly increased after the first increase in peroxidase concentration in a period of 2 h. There was no significant change on dye decolorization either on increasing peroxidase concentration or the incubation time.

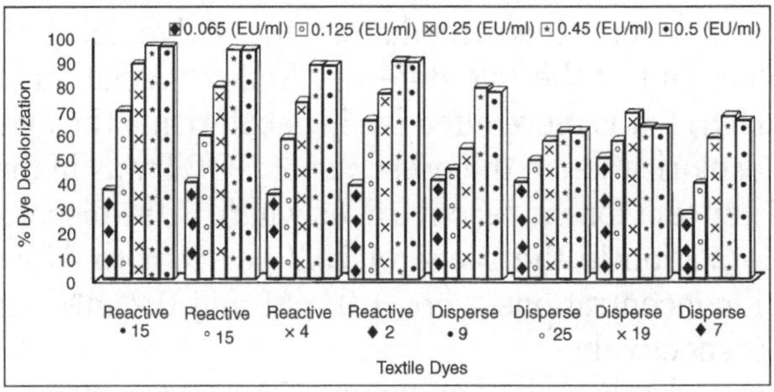

Fig. (a) Per cent Dye Decolorization of Textile Dyes in the Presence of Fixed Concentration of HOBT [0.2 mM for Reactive Dyes and 1.0 mM for Disperse Dye] and Increasing Concentration of *T. dioica* Peroxidase Enzyme (EU/ml). Please see Table-1 for λmax of each dye.

H_2O_2 and pH Activity Profile of Decolorization of Textile Dyes

Figure shows that the per cent decolorization improved with the increasing concentration of H_2O_2 and the maximum decolorization was observed at a concentration of 0.8 mM and 1.0 mM of H_2O_2 for disperse and reactive dyes respectively which remained substantially unaffected till 1.2 mM H_2O_2.

To find out the range of pH in which significant decolorization was observed; buffers in the range of pH 2.0 to pH 10.0 were used. The per cent decolorization is shown in. An acidic range of pH (3.0 to 6.0) was better suited for dye decolorization. Maximum decolorization was observed at pH 4.0 and pH 5.0 at fixed concentration of *T. dioica* peroxidase and HOBT for disperse and reactive dyes respectively. There was significant decrease in the extent of decolorization in an alkaline medium and at pH 10.0 the decolorization action of the enzyme was almost insignificant/ lost.

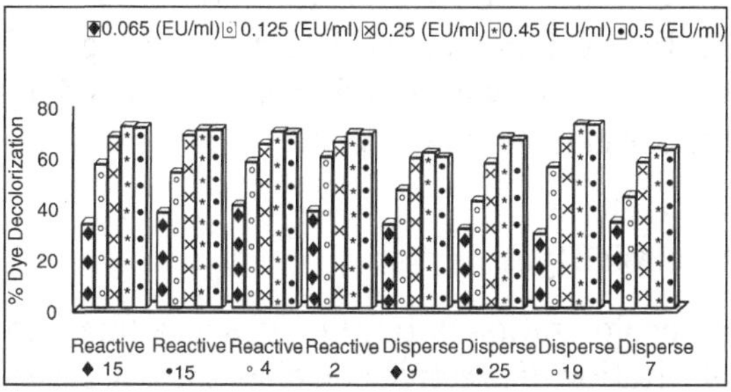

Fig. (b) Per cent Dye Decolorization of Textile Dyes in the Presence of Fixed Concentration of Vanillin [0.2 mM for Reactive Dyes and 1.0 mM for Disperse Dye] and Increasing Concentration of *T. dioica* Peroxidase Enzyme (EU/ml). Please see Table-1 for λmax of each dye.

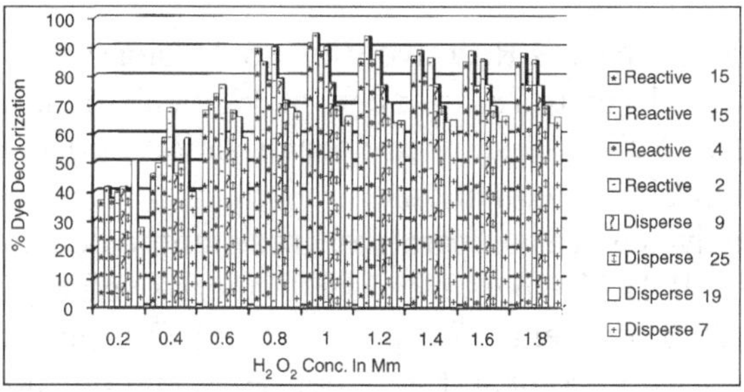

Fig. Per cent Dye Decolorization of Textile Dyes in the Presence of Fixed Concentration of HOBT, *T. dioica* Peroxidase Enzyme (EU/ml) and Varying Concentration of H_2O_2. Please see Table-1 for λmax of each dye.

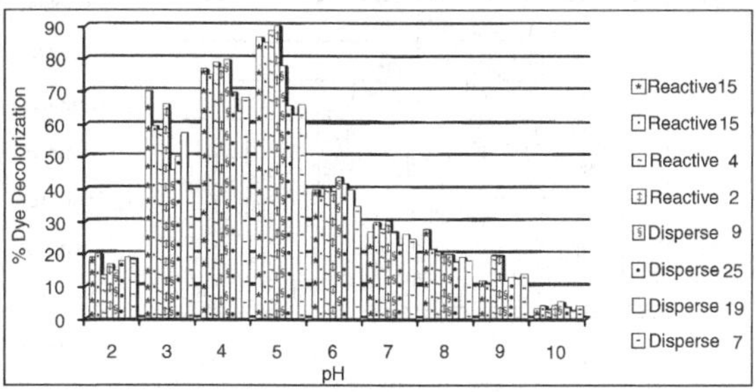

Fig. Per cent Dye Decolorization of Textile Dyes in the Presence of Fixed Concentration of HOBT, *T. dioica* Peroxidase Enzyme (EU/ml) and Varying pH. Please see Table-1 for λmax of each dye.

Temperature Activity Profile of Decolorization of Textile Dyes

The per cent decolorization was plotted as a function of temperature and the results are shown in. Among the textile dyes the reactive dyes exhibited maximum decolorization at 50°C whereas the disperse dyes showed maximum decolorization at 40°C in the presence of 1.0 mM and 0.2 mM HOBT.

Reactive Blue 15 (96.1 per cent), Reactive Orange 15 (94.4 per cent), Reactive Red 4 (85.2 per cent) whereas disperse dyes under study decolorized in the range of 61.2 per cent to 79 per cent.

Time Activity Profile of Decolorization of Textile Dyes

The extent of decolorization of textile dyes as a function of time is shown in. Maximum decolorization of reactive and disperse dyes were observed within 1 h of incubation at 50°C and 40°C with 0.45 EU/ml of PGP and 1.0 mM and 0.2 mM HOBT respectively.

However, further decolorization of these dyes progressed slowly upto 4 h, although no effective increase was observed even when the dyes were further incubated for longer times.

Among the reactive dyes Reactive Blue 15 decolorized almost completely and Reactive Orange 15 up to 78.3 per cent at 4 h incubation, whereas Reactive Red 4 and Reactive Yellow 2 showed decolorization up to 86.3 per cent and 81.2 per cent under similar conditions.

The disperse dyes were comparatively resistant to decolorization and only in the presence of HOBT Disperse Red 19, Disperse Yellow 7, Disperse Black 9 decolorized to 79.5 per cent, 72.3 per cent and 71.6 per cent respectively. Disperse Orange 15 was comparatively degraded and decolorized to a lesser extent under similar conditions with a maximum of 61.2 per cent.

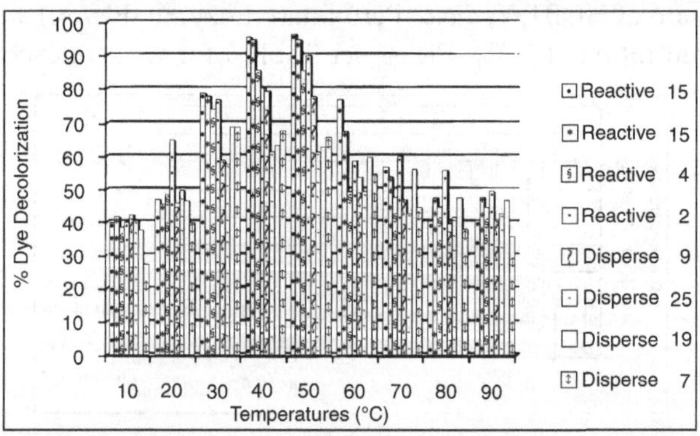

Fig. Per cent Dye Decolorization of Textile Dyes in the Presence of Fixed Concentration of HOBT, *T. dioica* Peroxidase Enzyme (EU/ml) at Different Temperatures. Please see Table-1 for λmax of each dye.

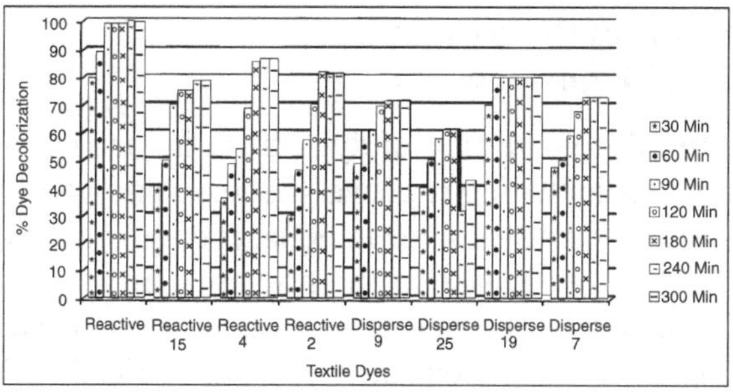

Fig. Per cent Dye Decolorization of Textile Dyes in the Presence of Fixed Concentration of HOBT, *T. dioica* Peroxidase Enzyme (EU/ml) at Different Time Interval. Please see Table-1 for λmax of each dye.

Decolorization Profile of Non-textile Dyesby *T. dioica* Peroxidase as a Function of Enzyme Concentration and Time

Nine different non-textile dyes were studied. These dyes were treated with different amount of *T. dioica* peroxidase in the range of 0.065 EU/ml to 0.50 EU/ml and incubated for varying time interval with and without the Redox mediator HOBT (1.0 mM) at 37°C as shown in. The results indicated that few non-textile dyes *viz.*, Methylene Blue, Martius Yellow, Methyl Orange, Rhodamine 6G were highly recalcitrant to decolorization by *T. dioica* in the absence of HOBT after 3h of incubation. However, the decolorization progressed slowly with the addition of 1.0 mM HOBT and per cent decolorization achieved for Methylene Blue, Martius Yellow and Methyl Orange, was 98.6 per cent, 35.1 per cent and 98.7 per cent respectively; whereas Rhodamine 6G almost decolorized completely. For other dyes, the maximum decolorization exhibited after 3h was 98.7 per cent, 97.3 per cent, 97.1 per cent, 67.8 per cent for Coomassie Brilliant Blue R 250, Naphthol Blue Black, Evans Blue and Eriochrome Black T respectively.

Table. Decolorization of Non Textile Dyes by *T. dioica* Peroxidase at Different Enzyme Concentration, Incubation Period and with Fixed Concentration of HOBT, H_2O_2.

Non Textile Dyes (λ max) nm	0.065 EU/ml		0.125 EU/ml		0.250 EU/ml		0.45 EU/ml		0.50Eu/ml	
	-HOBT (60min)	+ HOBT (60min)	- HOBT (120min)	+ HOBT (120min)	- HOBT (160min)	+ HOBT (160min)	- HOBT (180min)	+ HOBT (180min)	- HOBT (240min)	+ HOBT (240min)
Celestine (642)	46.6	57.3	49.7	68.7	54.3	69.3	55.6	79.5	58.9	79.6
Coomassie R250 (553 nm)	47.6	92.4	48.2	93.2	51.3	98.9	51.2	98.7	51.1	98.7
Methylene (664 nm)	2.5	67.2	3.7	78.5	3.6	98.4	3.5	98.6	3.2	98.9
Eriochrome (503 nm)	48.7	67.7	51.6	77.4	55.4	91.3	55.3	92.0	55.1	92.2
Evans (607 nm)	86.4	94.3	87.5	95.4	87.8	96.7	87.2	97.1	86.7	97.2
Martius (433 nm)	10.2	33.2	12.4	34.5	13.2	35.6	13.5	35.1	12.3	35.2
Methyl (464 nm)	12.3	87.2	13.2	89.2	13.2	98.5	13.1	98.7	12.2	98.6
Naphthol (618 nm)	47.2	78.8	49.2	86.2	56.6	97.4	57.2	97.3	57.9	97.1
Rhodamine (525 nm)	11.2	78.4	13.2	86.3	13.2	99.7	14.2	99.6	13.2	99.5

Decolorization Profile of Different Dye Mixtures by *T. dioica* Peroxidase

To simulate the decolorization of dyes from industrial effluent, complex mixtures of textile dyes including reactive, disperse and non-textile dyes were prepared by mixing four different dyes in equal proportions and incubated with 0.45 EU/ml of *T. dioica* peroxidase in the presence of 1.0 mM HOBT and 1.0 mM H_2O_2 for 2 h at 40°C. The decolorization was recorded at wave length maxima of each mixture determined spectrophotometrically. The combinations of different dyes showed decolorization by more than 82 per cent. The rate of decolorization of dye mixture was slower in comparison to that of individual dyes both in the presence and absence of HOBT. However, the HOBT mediated dye decolorization was more effective.

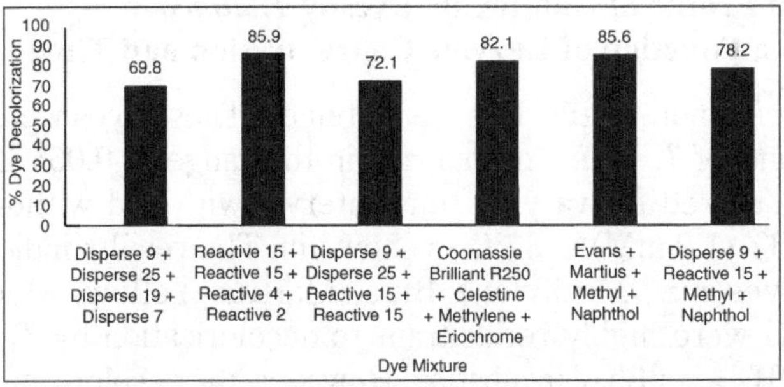

Fig. Degradation/ decolorization of dye mixture by *T. dioica* in the presence of HOBT/H_2O_2 to mimic dyes in industrial effluents. [Disperse 9+ Disperse 25 +Disperse 19+ Disperse 7 (λ_{461}); Reactive 15 +Reactive 15 + Reactive 4 + Reactive 2(λ_{531}); Disperse 9+ Disperse 25 + Reactive 15 +Reactive 15(λ_{521}); Coomassie Brilliant R250 + Celestine+ Methylene+ Eriochrome (λ_{595}); Evans+ Martius+ Methyl + Naphthol(λ_{541}); Disperse 9+ Reactive 15+ Methyl+ Naphthol (λ_{571})].

We have earlier reported the potential of a novel *T. dioica* plant proteins in the decolorization of disperse dyes (Jamal et al., 2010). In this paper we intended to widen the Spectrum of industrially important textile dyes and included some well known non-textile dyes to study the extent of decolorization in the presence of Redox mediators. The cost of an enzyme depends on its degree of purity hence we opted to use ammonium sulphate precipitated proteins from T. dioica to study dye decolorization/degradation. The PGP was partially purified using 10 per cent to 80 per cent ammonium sulphate which retained a specific activity of 96U/mg of protein. The experiments were performed at different enzyme concentration, pH, temperature, incubation period and Redox mediators namely, HOBT and vanillin. An interesting data profile was obtained for the assessment of this enzyme suitability to treat wastewater contaminated with these dyes.

The reactive dyes underwent decolorization by the formation of precipitate which disappeared in the presence of 1.0 mM HOBT. This finding supports earlier reports that treatment of phenols and aromatic amines by peroxidases resulted in formation of large insoluble aggregates (Wada et al., 1995; Tatsumi et al., 1996; Husain and Jan, 2000; Duran and Esposito, 2000). The other dyes studied showed no formation of precipitate during decolorization in the presence of HOBT. This observation supports earlier view that decolorization of dyes took place via degradation of aromatic ring of the compounds or by cleaving certain functional groups (Akhtar et al, 2005a,b; Satar and Husain, 2009). All the other textile dyes showed insignificant or no decolorization with T. dioica preoxidase alone when studied under optimum conditions of dye decolorization in the absence of HOBT (data not shown).

The decolorization profile for all the dyes increased significantly at HOBT concentration of 0.2 mM. Upon increasing the HOBT concentration further to 1.5 mM decolorization increased marginally. The reactive dyes under study exhibited decolorization maximally at 1.0 mM HOBT whereas disperse dyes showed maximum decolorization at lower values of HOBT suggesting that reactive dyes are comparatively more resistant to degradation. Vanullin was able to decolorize both the reactive and disperse dyes at 1.0 mM concentration but the extent of decolorization was sufficiently poor in comparison to HOBT for certain dyes like Reactive blue 15, Reactive orange 15 and most of the disperse dyes presence of relatively higher concentration of vanillin.

The non textile dyes exhibited remarkable decolorization in presence of both the redox mediators at 1.0 mM concentration. There was no significant change on dye decolorization of non- textile dyes at concentrations above 1.0 mM. It has already been reported that Redox mediators have the potential to mediate oxidation reaction between a substrate and an enzyme (d'Acunzo et al., 2006). Different Redox mediators have different mediation efficiency which is governed by Redox potential of the mediator and the oxidation mechanism of the substrate (Baiocco et al., 2003). Oxidation of substrate occurs by free radical formation by the mediator. The free radicals can be formed either by one-electorn oxidation of substrate or by abstraction of a proton from the substrate (Xu et al., 2001; Fabbrini et al., 2002). In this study, Redox-mediating property of two different compounds as peroxidase mediators was evaluated and extensive study on textile, non-textile and dye mixtures has been performed with HOBT.

Laccase have been used with Redox mediators to oxidize non-phenolic compounds (Bourbonnais and Paice, 1990). The mechanism of action of laccase mediator system has been extensively studied and it is used in the textile industry in the finishing process for indigo stained materials. Several workers have demonstrated that the use of Redox mediator system enhanced the rate of dye decolorization by several folds but these mediators were required in very high concentrations (Soares et al., 2001a,b; Claus et al., 2002). In the present study we have shown the decolorization of both textile and non-textile dyes as well as dye mixtures mediated by *T. dioica* peroxidase under the influence of low concentration of Redox mediators. The peroxidase reacted well to decolorize at concentration of 0.8 mM and 1.0 mM of H_2O_2 for disperse and reactive dyes respectively and remained substantially unaffected till 1.2 mM. Although the concentration of H_2O_2 greater than 0.75 mM acted as an inhibitor of peroxidase activity by irreversibly oxidizing the enzyme ferri-heme group essential for peroxidase activity (Satar and Husain, 2009), in this study we observed a relatively higher working concentration of H_2O_2. The enzyme in the presence of Redox mediators was much effective in performing the decolorization of the dyes, implying dye decolorization was a result of Redox mediated H_2O_2-dependent enzymatic reaction. Our results are consistent and very near to values reported earlier for maximum functional concentration of H_2O_2 (Camarero et al., 2005).The enzyme could function better in an acidic medium of pH range 3-6, whereas its decolorizing/degrading activity was adversely affected in alkaline medium. This finding supports the earlier view of disperse dye decolorization in acidic medium by *T. dioica* and reports further that acid medium favours catalysis of reactive dyes too by this peroxidase. It has earlier been reported that the degradation of industrially important dyes by enzymes such as horse radish peroxidase, polyphenol oxidase, BGP and laccase was also maximum in the buffers of acidic pH (Galindo et al., 2000).

The reaction temperature is an important parameter which effects the decolorization of dyes. The maximum decolorization for both the reactive and disperse dyes were in the temperature range of 40°C to 50°C in the presence of fixed concentration of HOBT. The extent of decolorization was remarkable for Reactive Blue 15, Reactive Orange 15, Reactive Red 4 at 50°C whereas disperse dyes could be decolorized in the range of 61.2 per cent to 79 per cent at 40°C. The decolorization varies with the nature of the dyes but Redox mediated decolorization with *T. dioica* peroxidase was a better solution for effective decolorization of recalcitrant compounds. The rate of decolorization varied with time and maximum decolorization in the presence of HOBT was observed within one hour of incubation for Reactive Blue15 and Disperse Red 19. However, for other dyes the extent of decolorization progressed slowly and reached a plateau after 4 h of incubation. This data is consistent with reports that decolorization rate varies, depending upon the type of dye to be treated (Camarero et al., 2005).

The non-textile dyes studied for decolorization with *T. dioica* peroxidase exhibited enhanced decolorization in the presence of 1.0 mM HOBT, whereas in the absence of Redox mediator decolorization was much slower. Coomassie Brilliant Blue R250, Methylene Blue, Eriochrome Black T, Martius Yellow, Methyl Orange, Naphthol Blue Black and Rhodamine were extensively decolorized by *T. dioica* peroxidase under the influence HOBT.The decolorization of Evans blue was not significantly affected by the presence of

Redox mediator, although in the presence of HOBT decolorization achieved was higher. The performance of this system was maximal during 160 min of incubation. Celestine blue was decolorized to 79.5 per cent in the presence of HOBT at higher concentration of enzyme as well as longer incubation time. The data in is suggestive of *T. dioica* peroxidase in conjunction with low concentration of HOBT to be wonderful decolorization/degradation system for non-textile dyes as well. Further, dye mixtures simulating industrial effluents exhibited more than 82 per cent decolorization with 1-hydroxybenzotriazole. The rate of decolorization of dye mixture was slower in comparison to that of individual dyes both in the presence and absence of HOBT. The application of inexpensive peroxidase from easily available source can overcome the limitations in current wastewater treatment strategies. The use of peroxidases can be extended to large-scale treatment of wide spectrum of structural dyes by using immobilized PGP along with relatively cheaper Redox mediators. Thus, the study demonstrated that the peroxidase enzyme isolated from *T. dioica* can be coupled with Redox mediator into a system that can serve as an effective biocatalyst for the treatment of effluents containing recalcitrant dyes from textile, dyeing and printing industries.

FUTURE PERSPECTIVE

Dye wastewater discharged from textile and dyestuff industries needs to be treated due to their impact on water bodies and to address growing public concern over their toxicity and carcinogenicity. Many different and complicated molecular structures of dyes make dye wastewater treatment difficult by conventional biological and physico-chemical processes. Therefore innovative treatment technologies need to be investigated. The studies performed using peroxidases from different sources indicates that novel enzyme systems can be created to decolorize wide spectrum of textile, non-textile dyes and dye mixtures under varying set of conditions. The efficacy of decolorization drastically improves with Redox mediators and dyes/dye mixtures recalcitrant to peroxidase exhibited remarkable decolorization. The application of inexpensive peroxidases from easily available sources can overcome the limitations in current wastewater treatment strategies. The use of peroxidases can be extended to large-scale treatment of wide spectrum of structural dyes by using immobilized peroxidases along with relatively cheaper Redox mediators.

4

Control Methods used in Treatment of Textile Wastewater

INTRODUCTION

Textile industry is one of the most water and chemical intensive industries worldwide due to the fact that 200-400 liters of water are needed to produce 1 kg of textile fabric in textile factories (Correia et al., 1994 and Orhon et al., 2003). The water used in this industry is almost entirely discharged as waste. Moreover, the loss of dye in the effluents of textile industry can reach up to 75 per cent (Couto and Toca-Herrera, 2006). The effluents are considered very complex since they contain salt, surfactants, ionic metals and their metal complexes, toxic organic chemicals, biocides and toxic anions. Azo dyes are regarded as the largest class of synthetic. Approximately, 50–70 per cent of the available dyes for commercial applications are azo dyes followed by the anthraquinone group (Konstantinou and Albanis, 2004). Azo dyes are classified according to the presence of azo bonds (–N=N–) in the molecule *i.e.*, monoazo, diazo, triazo etc. and also sub-classified according to the structure and method of applications such as acid, basic, direct, disperse, azoic and pigments (Bhutani, 2008). Some azo dyes and their dye precursors are well-known of high toxicity and suspected to be human carcinogens as they form toxic aromatic amines (Gomes et al., 2003 and Stylidi et al., 2003). Different physical, chemical and biological as well as the various combinations of pretreatment and post-treatment techniques have been developed over the last two decades for industrial wastewaters treatment in order to meet the ever-increasing requirements of human beings for water. Though there are numerous studies published in this field, most of the techniques adopted by these researchers are uneconomical, ineffective or impractical uses (Cooper, 1995 and Stephen, 1995). Recent studies have demonstrated that heterogeneous photocatalysis is the most efficient technique in the degradation of colored chemicals(Li et al., 2003; Vione et al., 2003; Antharjanam et al., 2003; Fernandez-Ibanez et al., 2003; Liu et al., 2003; Ohno, 2004; Chen et al., 2004; Alkhateeb et al., 2005 and

Attia et al., 2008). These studies used titanium dioxide and/ or zinc oxide in the photolysis processes. The large bang gap of titanium dioxide and zinc oxide (~ 3.2 eV) put a limitation of using these semiconductors in photocatalytic degradation under natural weathering conditions. Only a small part of the overall solar intensity could be useful in such photodegradation processes. However, the existence of dye on the surface of catalyst reduces the energy required for excitation and then increases the efficiency of the excitation process by extending its absorption in the visible region of the spectrum.

There are several factors to choose the appropriate textile wastewater treatment method such as, economic efficiency, treatment efficiency, type of dye, concentration of dye and environmental fate. There is no general method for the treatment of textile industrial wastewater. Wastewaters from textile industry contain various pollutants resulting from various stages of production, such as, fibres preparation, yarn, thread, webbing, dyeing and finishing. Mainly three methods are used for the treatment of textile industrial wastewater.

These are:

1. Physico-chemical methods.
2. Advanced oxidation methods.
3. Biological sludge methods.

The main operational methods used for the treatment of textile industrial water involve physical and chemical processes (Shaw et al., 2002 and Liu et al., 2005). However, these techniques have many disadvantages. These disadvantages include Sludge generation, high cost, formation of bi – products, releasing of toxic molecules, requiring a lot of dissolved oxygen, limitation of activity for specific dyes and requiring of long time. In recent years, advanced oxidation processes (AOPs) have gained more attraction as a powerful technique in photocatalytic degradation of textile industrial wastewater since they are able to deal with the problem of dye destruction in aqueous systems (Konstantinou and Albanis, 2004).

PHOTOCATALYSIS

Photocatalysis is defined as the acceleration of a photoreaction in the presence of a catalyst, while photolysis is defined as a chemical reaction in which a chemical compound is broken down by photons. In catalyzed photolysis, light is absorbed by an adsorbed substrate. Photocatalysis on semiconducting oxides relies on the absorption of photons with energy equal to or greater than the band gap of the oxide, so that electrons are promoted from the valence band to the conduction band:

$$\text{Semiconductor} + hv \rightarrow h^{+} + e^{-}$$

If the photoholes and photoelectrons produced by this process migrate to the surface, they may interact with adsorbed species in the elementary steps, which collectively constitute photocatalysis. The numerous gas- and liquid-phase reactions photocatalysed by TiO_2 and ZnO have been reviewed. There is a general agreement that adsorption is necessary since the surface species act as traps for both; photogenerated holes and electrons, which otherwise recombine. In addition to participating in conventional surface reaction steps, adsorbed dye molecules assist in the separation of photoholes and photoelectrons, which may otherwise recombine within the semiconductor particles. A major factor affecting the efficiency of photocatalysis process is electron/hole recombination. If the electrons and holes are used in a reaction, a steady state will be reached when the removal of electrons and holes equals the rate of generation by illumination. Recombination and trapping processes are the de-excitation processes which are responsible for the creation of the steady state, if no reaction occurs.

There are three important mechanisms of recombination:

1. Direct recombination.
2. Recombination at recombination centers.
3. Surface recombination.

There are different types of semiconductors, whose band gaps range between 1.4 and 3.9 eV, *i.e.*, it could be excited with a light of 318–886 nm wavelengths. This means that most of the known semiconductors could be excited by using visible light. However, not all these semiconductors could be used in the photocatalytic reactions. Bahnemann et al., (1994) report that the most appropriate photocatalysts should be stable towards chemicals and illumination and devoid of any toxic constituents, especially for those used in environmental studies. The authors also explaine that TiO_2 and ZnO are the most commonly used in photocatalytic reactions due to their efficient absorbtion of long wavelength radiation as well as their stability towards chemicals. Other semiconductors like WO_3, CdS, GaP, CdSe and GaAs absorb a wide range of the solar spectrum and can form chemically activated surface-bound intermediates, but unfortunately these photocatalysts are degraded during the repeated catalytic cycles involved in heterogeneous photocatalysis.

PHOTOSENSITIZATION

The illumination of suspended semiconductor in an aqueous solution of dye with unfiltered light (polychromatic light) leads to the possibility of the existence of two pathways (Hussein et al., 2008):

1. In the first pathway, the part of light with energy equal to or more than the band gap of the illuminated semiconductor will cause a promotion of an electron to conduction band of the semiconductor and as a result, a positive hole will be created in the valence band. The formed photoholes and photoelectrons can move to the surface of the semiconductor in the presence of light energy. The positive hole will react with adsorbed water molecules on the surface of

semiconductor producing $^{\bullet}OH$ radicals and the electron will react with adsorbed oxygen on the surface. Moreover, they can react with deliquescent oxygen and water in suspended liquid and produce perhydroxyl radicals ($HO_2^{\bullet}$) with high chemical activity (Zhao and Zhang, 2008). The processes in this pathway can be summarized by the following equations:

$$h^+ + OH^- \rightarrow \dot{O}H$$

$$h^+ + H_2O \rightarrow H^+ + \dot{O}H$$

$$e^- + O_2 \rightarrow O_2^{\bullet -}$$

$$O_2^{\bullet -} + H^+ \rightarrow HO_2^{\bullet}$$

$$HO_2^{\bullet} + O_2^{\bullet -} \xrightarrow{H^+} H_2O_2 + O_2$$

$$H_2O_2 \rightarrow 2\dot{O}H$$

$$\text{Dye} + \text{Semiconductor}\ (h^+_{VB}) \rightarrow \text{Dye}\cdot^{+} + \text{Semiconductor}$$

$$\text{Dye}\cdot^{+} + O\cdot_2^{-} \rightarrow DO_2 \rightarrow \text{degradation products}$$

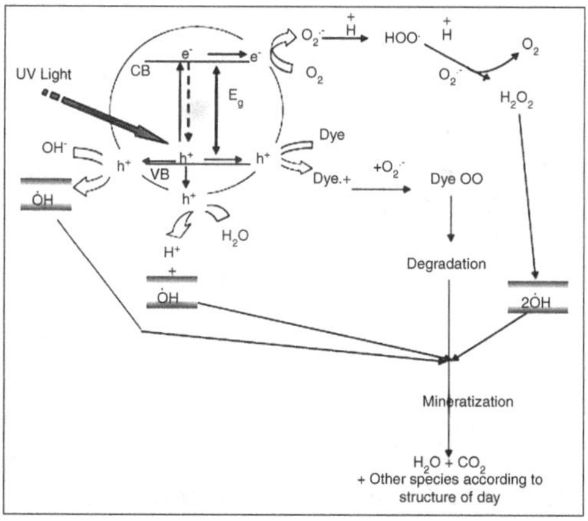

Fig. Dye/Semiconductor/UV Light System

2. In the second pathway, the other part of light with energy which is less than the band gap of the illuminated semiconductor will be absorbed by the adsorbed dye molecules. Dye molecules will be decolorized by a photosensitization process. The photocatalytic decolorization of dyes, which is described as photosensitization processes, is also characterized by a free radical mechanism. In this process, the adsorbed dyes molecules on the surface of the semiconductor can absorb a radiation in the visible range in addition to the radiation with a short wavelengths (Fernandez-Ibanez et al.,

2003; Ohno, 2004 and Alkhateeb et al., 2005). The excited colored dye (dye*) (in the singlet or triplet state) will inject an electron to the conduction band of the semiconductor (Hussein and Alkhateeb, 2007). The processes in this pathway can be summarized by the following equations:

$$\text{Dye} + \text{hv}\left(\text{VIS or UV reagion}\right) \rightarrow {}^{1}\text{Dye}^{*} \text{or } {}^{3}\text{Dye}^{*}$$

$${}^{1}\text{Dye}^{*}\text{or } {}^{3}\text{Dye}^{*} + \text{semiconducto} \rightarrow \text{Dye}^{\cdot+} + e^{-}\left(\text{to the conduction band of semiconductor}\right)$$

$$e^{-} + O_{2} \rightarrow O_{2}^{\bullet-} + \text{Semiconductor}$$

$$\text{Dye}^{\cdot+} + \text{O}\cdot_{2}^{-} \rightarrow \text{Dye O}_2 \rightarrow \textit{degradation} \text{ products}$$

$$\text{Dye}^{\cdot+} + \text{HO}\cdot_{2} \text{ (orHO)} \rightarrow \textit{degradation} \text{ products}$$

$$\text{Dye} + 2\dot{\text{O}}\text{H} \rightarrow \text{H}_2\text{O} + \text{oxidation products}$$

$$\text{Dye}^{\cdot+} + \bar{\text{O}}\text{H} \rightarrow \text{Dye} + \dot{\text{O}}\text{H}$$

$$\text{Dye}^{\cdot+} + \text{H}_2\text{O} \rightarrow \text{Dye} + \dot{\text{O}}\text{H} + \text{H}^{+}$$

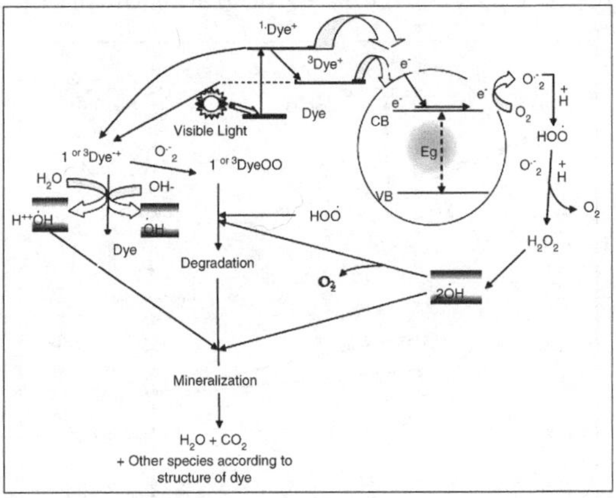

Fig. Nonregenerative Dye/Semiconductor/Visible Light System

The mechanism above is favoured by non-regenerative organic dye where dye/semiconductor/visible light system and the sensitizer itself degrade. However, in regenerative semiconductor system, the following mechanism may be followed:

$$\text{Sen} + \text{TiO}_2 \leftrightarrow \text{Sen} - \text{TiO}_2$$

$$\text{Sen - TiO}_2 + \text{RX} \leftrightarrow \text{Sen} - \text{TiO}_2\text{......RX}_{(\text{ads})}$$

$$\text{Sen} - \text{TiO}_2 + \text{O}_2 \leftrightarrow \text{Sen} - \text{TiO}_2\text{......O}_{2(\text{ads})}$$

$$\text{Sen} - \text{TiO}_2 \underset{\text{hv}}{\overset{\text{hv visible light}}{\rightleftarrows}} \text{Sen}^{*} - \text{TiO}_2$$

$$Sen^* - TiO_2......O_{2(ads)} \rightarrow Sen^+ - TiO_2 + O^-_2$$

$$Sen^+ - TiO_2 + O_2^- \rightarrow Sen - TiO_2 + O_2$$

$$Sen^* - TiO_2\ Sen^+ - TiO_2\ (e^-_{C.B})$$

$$Sen^+ - TiO_2\ (e^-_{C.B}) \rightarrow Sen - TiO_2$$

$$Sen^+ - TiO_2\left(e^-_{C.B}\right)......RX \rightarrow Sen^+ - TiO_2 + \dot{R}X + X^-$$

$$Sen^+ - TiO_2\left(e^-_{C.B}\right)......O_{2(ads)} \rightarrow Sen^+ - TiO_2 + O_2^-$$

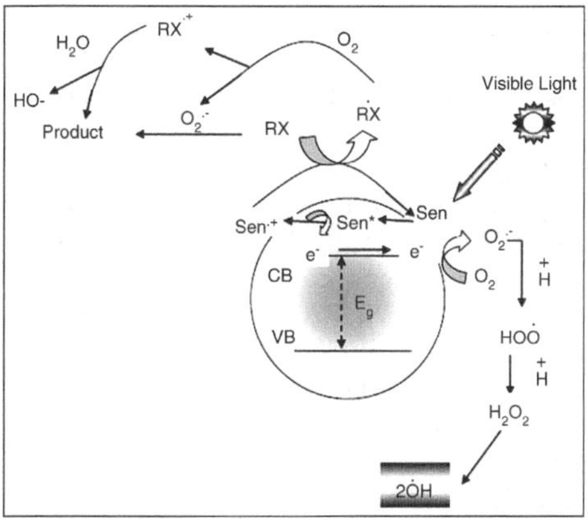

Fig. Regenerative Dye/Semiconductor/Visible Light System

Cho et al., (2001) conclude that there is no direct electron transfer between an excited sensitizer and CCl_4 molecules, in homogeneous solution, and the existence of semiconductor is essential for sensitized photocatalysis. Platinum supported on titanium dioxide acts as an excellent sensitizer and could have practical advantages as a mild and convenient photocatalyst for selective oxidation processes (Hussein et al., 1984). The addition of Rhodamine-B, as sensitizer to TiO_2 dispersion system increases the rate of photooxidation properties (Hussein et al., 1991). The authors explained that due to the fact that more light absorbed by Rhodamine-B between 460-580 nm, then the energy transfer from sensitizer to TiO_2 or to any other active species and hence promote the photocatalytic activity of titanium dioxide.

ADVANCED OXIDATION PROCESSES

Glaze et al., (1987) define Advanced Oxidation Processes (AOPs) for water treatments as the processes that occur near ambient temperature and pressure which involve the generation of highly reactive radicals,especially hydroxyl radicals (•OH), in sufficient quantity for water purification. Advanced oxidation processes can also be easily defined as techniques of destruction of organic pollutants from wastewaters. These processes include chemical oxidation processes using hydrogen peroxide, ozone, combined ozone and hydrogen peroxide, hypochlorite, Fenton's reagent, ultra-violet enhanced oxidation such as UV/O_3, UV/ H_2O_2, UV/air, wet air oxidation and catalytic wet air. Hydroxyl radicals are strong reactive species, which are capable of destroying a wide range of organic pollutants. Table shows hydroxyl radical as the second strongest oxidant (Weast, 1977; Legrini et al., 1993 Domènech et al., 2001; and Mota et al., 2008).

Table. Standard Reduction Potential of Common Oxidants against Standard Hydrogen Electrode

Oxidant	E° (V)
Fluorine (F_2)	3.03
Hydroxyl radical (•OH)	2.80
Atomic oxygen (O)	2.42
Ozone (O_3)	2.07
Hydrogen peroxide(H_2O_2)	1.78
Hydroperoxyl radical ($O_2H\cdot$)	1.70
Potassium permanganate ($KMnO_4$)	1.67
Hypobromous acid (HBrO)	1.59
Chlorine dioxide (ClO_2)	1.50
Hypochlorous acid (HClO)	1.49
Hypochloric acid	1.45
Chlorine (Cl_2)	1.36
Bromine (Br_2)	1.09
Iodine (I_2)	0.54

The attack of organic pollutants by hydroxyl radicals occurs via the following mechanisms (Buxton et al., 1988; Legrini et al., 1993 and Pignatello et al., 2006):

- Electron transfer from organic pollutants to hydroxyl radicals:

$$\dot{O}H + RX \rightarrow RX^{\cdot +} + OH^{-}$$

- Hydrogen atom abstraction from the C–H, N–H or O–H bonds of organic pollutants:

$$\dot{O}H + RH \rightarrow R^{\cdot} + H_2O$$

- Addition of hydroxyl radical to one atom of a multiple atom compound:

$$\dot{O}H + Ph \rightarrow HOPh^{\cdot}$$

FUNDAMENTAL PARAMETERS IN PHOTOCATALYSIS

In semiconductor photocatalysis of industrial wastewater treatment, there are different parameters affecting the efficiency of treatment. These parameters include mass of catalyst, dye concentration, pH, light intensity, addition of oxidizing agent, temperature and type of photocatalyst. Other factors, such as, ionic components in water, solvent types, mode of catalyst application and calcinations temperature can also play an important role on the photocatalytic degradation of organic compounds in water environment (Guillard et al., 2005 and Ahmed et al., 2011).

EFFECT OF TYPE OF CATALYST

Haque and Muneer (2007) observe that Degussa P25 is more reactive for degradation of a textile dye derivative, bromothymol blue, in aqueous suspensions than other commercially available photocatalysts types of titanium dioxide, namely Hombikat UV100, PC500 and TTP. They explain the high activity of Degussa P25 is due to composing of small nano-crystallites of rutile dispersed within the anatase matrix. The band gap of rutile is less than that of anatase and as a result electron will transfer from the rutile conduction band to electron traps in anatase and the recombination of electrons and holes will be reduced. Hussein, (2002) reported that anatase has higher photoactivity than rutile due to the difference in surface area. Decolorization percentage of real textile industrial wastewater on rutile, anatase, and zinc oxide shows that the activity of different catalysts falls in the following sequence (Hussein and Abass, 2010 a):

$$ZnO > TiO_2 \text{ (Anatase)} > TiO_2 \text{ (Rutile)}$$

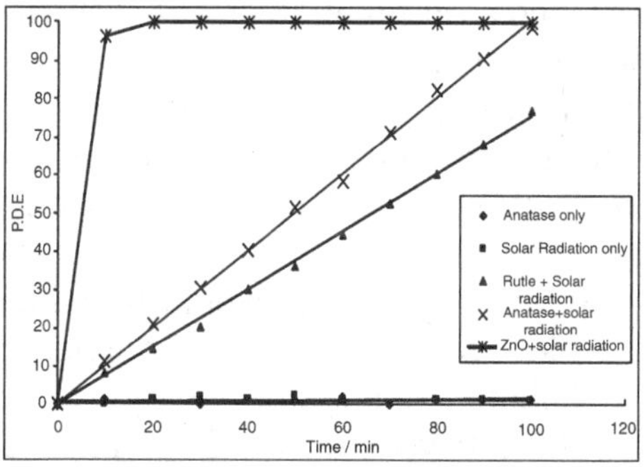

Fig. Photocatalytic Decolorization of Real Textile Industrial Wastewater at Different Conditions.

ZnO is more active than TiO_2 due to the absorption of wider spectrum light (Sakthivela et al., 2003). However, the amount of zinc oxide required to reach the optimum activity is two times more than that for titanium dioxide (anatase or rutile) (Hussein and Abass, 2010 a). In another study, Hussein et al., (2008) observed that ZnO is less active than anatase when the same weight of catalysts is used for photocatalytic degradation of textile wastewater. Akyol et al., (2004) reported that ZnO is more active than TiO_2 for the decolorization efficiency of aqueous solution of a commercial textile dye due to the band gap energy, the charge carrier density, and the crystal structure. Decolorization efficiency of real textile industrial wastewater in the presence and absence of catalyst and/or solar radiation was also investigated (Hussein and Abass, 2010 b). The results indicate that the activity of different catalysts fall in the sequence:

$ZnO > TiO_2$ (Anatase) $> TiO_2$ (Rutile) > in the absence of catalyst = in the absence of solar radiation or artificial radiation = 0

The results are plotted in Figure. These results also indicate that there has been no dark reaction. Incubations of colored industrial wastewater without solar radiation and/or without catalyst has been performed to demonstrate that decolorization of the dye is dependent on the presence of both; light and catalyst.

EFFECT OF MASS OF CATALYST

Photocatalysts dosage added to the reaction vessel is a major parameter affecting the photocatalytic degradation efficiency (Dong et al., 2010). Photocatalytic degradation efficiency increases with an increase in catalysts mass. This behaviour may be due to an increase in the amount of active site on surface of photocatalyst particles. As a result, an increasing the number of dye molecule adsorbed on the surface of photocatalyst lead to an increase in the density of particles in the area of illumination (Kim and Lee, 2010). The extrapolation of Hird's data (Hird, 1976) indicates that only 7.5 mg of TiO_2 was sufficient to absorb all incident 366 nm radiation. It follows that the mass effect must be caused by changes in the effective utilization of the absorbed radiation rather than by the increased absorption. Photocatalyst with small particles are more efficient than larger particles.

This behaviour may be due to (Hussein, 1984):

- Photoholes and photoelectrons generated in the bulk would have fewer traps and recombination centers to overcome before reaching the surface.
- A greater proportion of material would be within the space charge arising from depletive oxygen chemisorptions, which favour exciton dissociation and photohole migration to the surface.

Hence, increasing the catalyst's mass will increase the concentration of the efficient small particles within the illuminated region of the reaction vessel. The direct proportionality between photocatalytic degradation efficiency and catalyst loading is real within low concentrations of photocatalyst where there are excess active sites reaching plateau reign. The plateau is reached when this effect can no longer increase the overall efficiency of utilizing incident radiation. Moreover, after the plateau region is achieved, the activity of photocatalytic decolorization decrease with increase of catalyst concentration for all types of catalysts. This behaviour is more likely to emanate from variation in the intensity of radiation entering the reaction vessel and the way the catalyst utilizes that radiation. Light scattering by catalyst particles at higher concentration lead to decrease in the passage of irradiation through the sample leading to poor light utilization (Gaya et al., 2010; Kavitha and Palanisamy, 2011). Deactivation of activated photocatalyst molecules colliding ground state molecules with increasing the load of photocatalyst may be also cause reduction in photocatalyst activity (Kim and Lee, 2010). Photocatalytic decolorization efficiency (PDE) per cent of real textile industrial wastewater has been investigated by employing different masses of TiO_2 (anatase or rutile) or ZnO under natural weathering conditions for 20 minutes of irradiation (Hussein and Abass, 2010 b). The results are plotted in Figure.

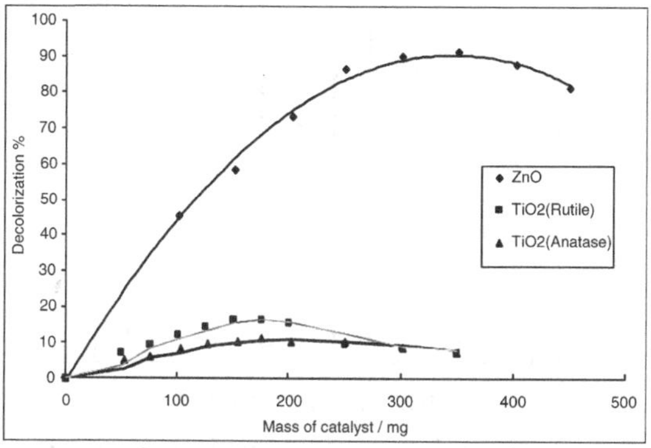

Fig. Effect of Mass on Photocatalytic Decolorizat on Efficiency of Real Textile Industrial Wastewater

The results in all cases indicate that the decolorization efficiency increases with increase in catalysts mass and then it becomes constant. It is clear from consideration of the catalyst concentrations at which the activity plateau were achieved that the mass effect does not depend upon the type of dye and source of irradiation. Moreover, plateau regions were achieved and then the activity of decolorization decreased with increasing catalyst concentration, for all types of catalysts used in this project.

EFFECT OF PH

Aqueous solution pH is an important variable in the evaluation of aqueous phase mediated photocatalytic decolorization reactions. pH change effects the adsorption quantity of organic pollutants and the ways of adsorption on the surface of photocatalyst (coordination).

As a result, the photocatalytic degradation efficiency will greatly be influenced by pH changes. Zero Point Charge (pHzpc), is a concept relating to adsorption phenomenon and defined as the pH at which the surface of an oxide is uncharged.

If positive and negative charges are both present in equal amounts, then this is the isoelectric point (iep).

However, the zpc is the same as iep when there is no adsorption of other ions than the potential determining H^+/OH^- at the surface. In aqueous solution, at pH higher than pHzpc, the oxide surface is negatively charged and then the adsorption of cations is favoured and as a consequence, oxidation of cationic electron donors and acceptors are favoured.

At pH lower than pHzpc, the adsorbent surface is positively charged and then the adsorption of anions is favoured and as a consequence, the acidic water donates more protons than hydroxide groups.

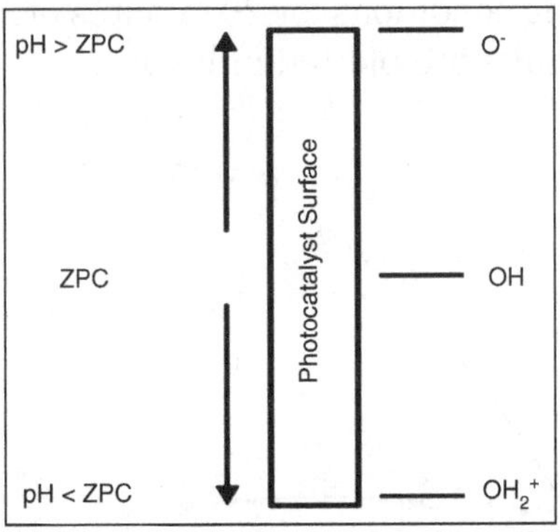

Fig. Effect of pH on ZPC.

Infrared spectroscopy study of Szczepantiewicz et al., (2000) shows that the TiOH sites are the major electron traps when TiO_2 is illuminated. The distribution of other species ($TiOH_2^+$ and TiO^-) with changing pH has been proposed by Kormann and co-workers (1991). Figures show that, at pH below ZPC the surface is mostly positively charged and TiOH sites increase as pH increases and reach maximum value at ZPC of semiconductor. However, $TiOH_2^+$ as pH increases and reaches zero value at ZPC. At pH higher than ZPC the density of TiO- groups on the surface start to form and reached 100 per cent value at pH 14. The importance of pH during the reaction is not less than that of initial state. The formation of intermediate products, sometimes, changes the pH of aqueous solution and as a result, it affects the rate of photodegeradation (Galvez, 2003).

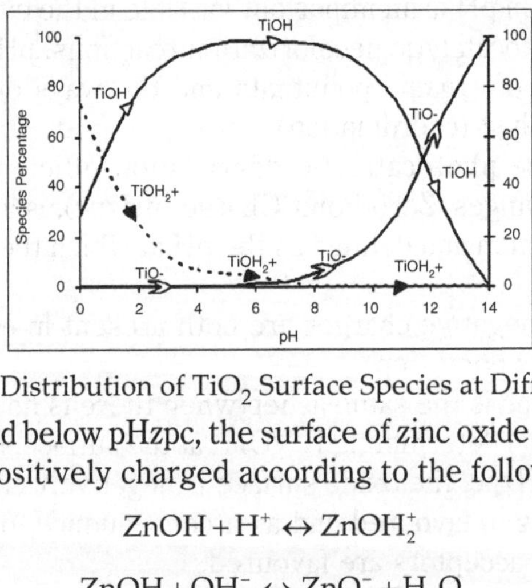

Fig. The Distribution of TiO_2 Surface Species at Different pH

At pH above and below pHzpc, the surface of zinc oxide and titanium dioxide are negatively or positively charged according to the following equations:

$$ZnOH + H^+ \leftrightarrow ZnOH_2^+$$

$$ZnOH + OH^- \leftrightarrow ZnO^- + H_2O$$

$$TiOH + H^+ \leftrightarrow TiOH_2^+$$

$$TiOH + OH^- \leftrightarrow TiO^- + H_2O$$

Effect of pH on Photocatalytic Decolorization of Bismarck Brown R

Under the determined experimental condition with initial dye concentration equal to 10^{-4} M, ZnO dosage 3.75 gm.L^{-1}, light intensity equal to 2.93 mW.cm^{-2} and temperature equal to 298.15 K, the effect of change in solution pH on decolorization percentage has been studied in the range 2-12.

The decolorization per cent has been found to be strongly dependent on pH of solution because the reaction takes place on the surface of semiconductor. The decolorization percentage of Bismarck brown R increases with the increase of pH, exhibiting maximum decolorization at pH 9.

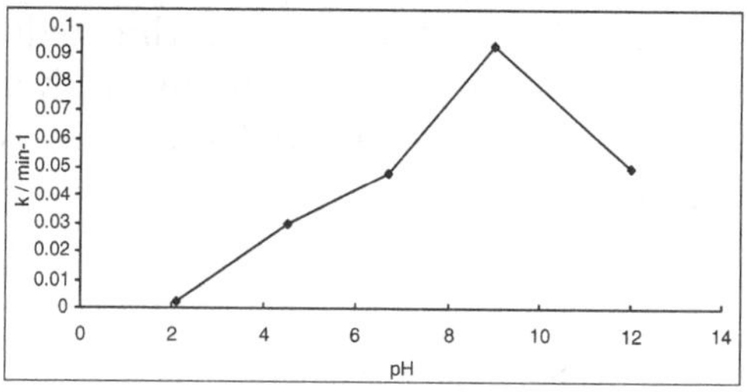

Fig. Effect of pH on Photocatalytic Decolorization Efficiency of Bismarck R Brown on ZnO

Under the determined experimental condition with initial dye concentration equal to 10^{-4}M, TiO_2 dosage 1.75 gm.L^{-1}, light intensity equal to 2.93 mW.cm^{-2} and temperature equal to 298.15 K, the effect of change in solution pH on decolorization percentage has been studied in the range 2-10. The results are plotted in Figures, for TiO_2 (DegussaP25), TiO_2 (HombikatUV100), TiO_2 (MillenniumPC105) and TiO_2 (Koronose2073).

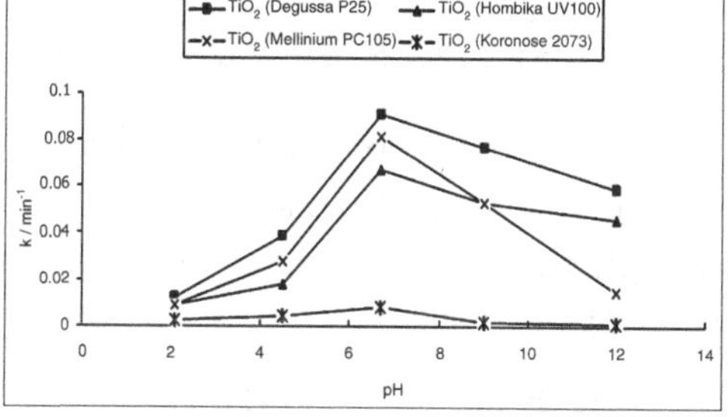

Fig. Effect of pH on Photocatalytic Decolorization Efficiency of Bismarck R Brown on Different Types of TiO_2

It was observed that the decolorization percentage strongly depends on the pH of solution because the reaction takes place on the surface of semiconductor. The decolorization percentage of Bismarck brown R increases with the increase of pH, exhibiting maximum decolorization at pH that is equal to 6.61, 6.54, 6.75, 6.63 for TiO2 (DegussaP25), TiO_2 (HombikatUV100), TiO_2 (MillenniumPC105) and TiO_2 (Koronose2073), respectively.

This behaviour could be explained (as mentioned before) on the basis of zero point charge (ZPC). The zero point charge is equal to 9.00 for ZnO and 6.25 for TiO_2 (Degussa P25). With the increase of the pH of solution, the surface of catalyst will be negatively charged by adsorbed hydroxyl ions

The presence of large quantities of adsorbed OH^- ions on the surface of catalyst favour the formation of •OH radical.

However, if pH is lower than ZPC, the hydroxyl ions adsorbed on the surface will be decreased and, therefore, hydrogen ions adsorbed on the surface will increase and the surface will become positive charged. Both the acidic and basic media leave an inverse impact on the photodecolorization efficiency because of the decrease of the formation of the hydroxyl radical. The decolorization of Bismarck brown R decreases dramatically at strong acid media (pH = 2.1) for ZnO. This could be explained due to photocorrosion of ZnO as shown in the following equations:

$$ZnO \xrightarrow{hv} e^-_{CB} + h^+_{VB}$$

$$ZnO + 2h^+_{VB} \rightarrow Zn^{2+} + \frac{1}{2}O_2$$

EFFECT OF LIGHT INTENSITY

Egerton and King, (1979) show that square root of light intensity depends on the activity of titanium dioxide for different wavelengths of light. However, this relationship cannot be applied to all range of light intensities. The primary electronic processes which occur in the absorption of photons with energy equal or greater than the band gap of semiconductor are:

$$\text{Semiconductor} + hv \xrightarrow{k_1} (h-e)\text{exciton}$$

$$(h-e) \xrightarrow{k_2} h^+ + e^-$$

$$h^+ + e^- \xrightarrow{k_3} \text{Radiationless Re combination}$$

For photocatalysis processes, it is necessary that the excitons dissociate and the photoholes and photoelectrons reach the catalyst surface where they are trapped by surface species:

$$h^+OH^-_{(S)} \xrightarrow{k_4} \dot{O}H_{(S)}$$

$$e^- + O_2 \xrightarrow{k_5} O^-_{2(ads)}$$

The concentration of excitons, photoholes and photoelectrons may be considered by applying a steady state treatment:

$$\frac{d[h-e]}{dt} = k_1 I_{(ads)} - k_2[(h-e)] = 0$$

So that:

$$[(h-e)] = \frac{k_1}{k_2} I_{(ads)}$$

Similarly:

$$\frac{d[h]}{dt} = k_2[(h-e)] - k_3[h^+][e^-] - k_4[h^+][OH_{(S)}] = 0$$

Since:

$$[h^+] = [e^-]$$

Then:

$$\frac{d[h]}{dt} = k_1 I_{(ads)} - k_3[h^+]^2 - k_4[h^+][OH^-_{(S)}] = 0$$

So that:

$$k_1 I_{(abs)} = k_3[h^+]^2 + k_4[h^+][OH^-_{(S)}]$$

There are two possibilities concerning the light intensity:

- At high light intensities, where the recombination of photoholes and photoelectrons is predominate, then:

$$k_3[h^+]^2 \rangle\rangle k_4[h^+][OH^-_{(S)}]$$

So equation 48 becomes:

$$k_1 I_{(abs)} = k_3[h^+]^2$$

Then:

$$[h^+] = \left[\frac{k_1}{k_3}\right]^{\frac{1}{2}} I^{\frac{1}{2}}_{(abs)}$$

If the rate controlling step in the overall photocatalysis processes involves the surface trapping of photoholes at surface OH-, then the rate will be given by equation 40.

The reaction rate is given by:

$$\text{Reaction rate} = k_4 \left(\frac{k_1}{k_3}\right)^{\frac{1}{2}} I^{\frac{1}{2}}_{(abs)} [OH^-_{(S)}]$$

Then:

$$\text{Reaction rate}\, \alpha [I_{(abs)}]^{\frac{1}{2}}$$

If, on the other hand, the rate controlling step involves photoelectron trapping by oxygen, then the rate controlling step will be equation 41. The reaction rate is given by:

$$\text{Reaction rate} = k_5 \left(\frac{k_1}{k_3}\right)^{\frac{1}{2}} I_{abs}\left[O_{2(abs)}\right]$$

Then:

$$\text{Reaction rate}\,\alpha\left[I_{(abs)}\right]^{\frac{1}{2}}$$

- At low light intensities, it is expected that recombination of photoholes and photoelectrons will be low, then:

$$k_4\left[h^+\right]\left[OH^-_{(S)}\right]\rangle\rangle k_3\left[h^+\right]^2$$

So equation 47 becomes:

$$k_1 I_{(abs)} = k_4\left[h^+\right]\left[OH^-_{(S)}\right]$$

Then:

$$\left[h^+\right] = \frac{k_1\left[I_{(abs)}\right]}{k_4\left[OH^-_{(S)}\right]}$$

It follows that:

$$\text{Reaction rate} = k_1\left[I_{(abs)}\right]$$

Alternatively, if photoelectron trapping is considered to be rate controlling, then:

$$\text{Reaction rate} = k_5 \frac{k_1}{k_4}\left[\left(I_{(abs)}\right)\right]$$

Hence, a linear dependence would be expected at low light intensities. Square-root intensity dependence was observed with rutile I, rutile 11, anatase, uncoated anatase pigment and platinized anatase and also independent on wavelength of incident radiation (Harvey et al., 1983 a; Hussein and Rudham, 1987). Bahnemann et al., (1991) reported that the change in kinetic constant is a function of the square root of the radiation entering at high light intensities, while this change can be linear with light intensity of incident radiation low light intensities (Peterson et al., 1991). Ollis et al., (1991) summarize the effect of light intensity on the kinetics of the photocatalytic degradation of dye as follows:

- At low light intensities (0–20 mW/cm^2), the rate of photocatalytic degradation is proportional directly with light intensity (first order).

- At high light intensities (25 mW/cm^2), the rate of photocatalytic degradation is proportional directly with the square root of the light intensity (half order).
- At high light intensities the rate of photocatalytic degradation is independent of light intensity (zero order).

However, Hussein et al (2011) found that the rate of photocatalytic decolorization of Bismarck brown R on ZnO and different types of titanium dioxide is proportional directly with the light intensity of incident UVA radiation in the range of 0- 2.0 mW/cm^2 and with the square root of light intensity in the range of 2.0- 3.5 mW/cm^2.

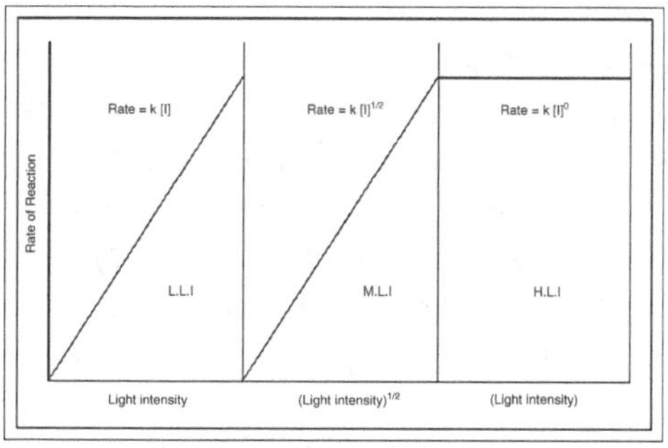

Fig. Effect of Light Intensity on the Kinetics of the Photocatalytic Degradation of Dye

Figures illustrate the impact of initial light intensity on the value of rate constant for photocatalytic decolorization of Bismarck brown R on ZnO and different types of titanium dioxide, respectively. The results indicate that the photocatalytic decolorization of Bismarck brown R increases with the increase in light intensity, attaining a maximum value at 3.52 mW.cm^{-2}.

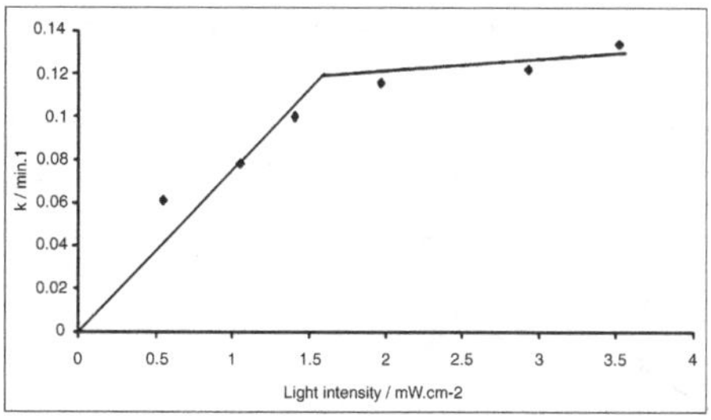

Fig. Effect of Initial Light Intensity on Rate Constant of Photocatalytic Decolorization of Bismarck Brown R on ZnO.

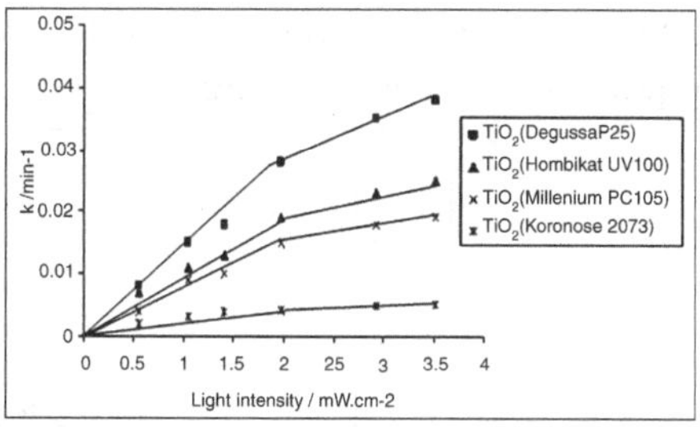

Fig. Effect of Initial Light Intensity on Rate Constant of Photocatalytic Decolorization of Bismarck Brown R using Different Types of TiO_2.

EFFECT OF TEMPERATURE

One of the advantages of photoreaction is that it is not affected or slightly affected by temperature change. Temperature dependent steps in photocatalytic reaction are adsorption and desorption of reactants and products on the surface of photocatalyst. None of these steps appears to be rate determining. The impact of temperature is explained as the variable with the least effect on photocatalytic degradation of aqueous solution of azo dyes (Obies, 2011). Attia et al., (2008) have found that the activation energy of photodegeradation of real textile industrial wastewater is equal to 21 ± 1 kJ mol^{-1} on titanium dioxide and 24 ± 1 kJ mol^{-1} on zinc oxide.

The activation energy for the photocatalytic degradation of textile industrial wastewater on titanium dioxide is similar to previous findings for photocatalytic oxidation of different types of alcohols on titanium dioxide and metalized titanium dioxide (Al-zahra et al., 2007; Hussein and Rudham, 1984, 1987). The single value of activation energy (21 ± 1 kJ mol^{-1}) that can be related to the calculated activation energy of photooxidation of different species of titanium oxide is associated with the transport of photoelectron through the catalyst to the adsorbed oxygen on the surface (Harvey et al., 1983 a and b).

Kim and Lee, (2010) explained that the very small activation energy in photocatalytic reactions is the apparent activation energy E_a, whereas the true activation energy E_t is nil. These types of reactions are operating at room temperature.

Palmer et al., (2002) observed that the effect of temperature on the photocatalytic degradation is insignificant in the range of 10-68°C. High temperatures may have a negative impact on the concentration of dissolved oxygen in the solution and consequently, the recombination of holes and electrons increases at the surface of photocatalyst.

However, Trillas et al., (1995); Chen and Ray, (1998) reported that raising the temperature of reaction enhances the rate of photocatalytic degradation significantly. Hussein and Abbas (2010 b) reported that the decolorization efficiency of real textile industrial wastewater increases with increasing of temperature as shown in fig. 13.

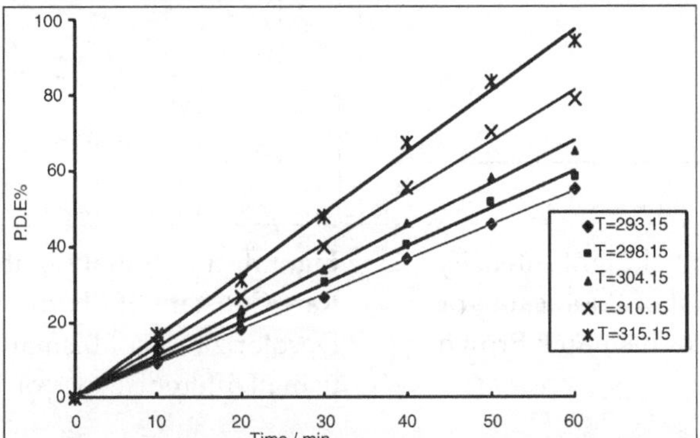

Fig. Effect of Temperature on P.D.E. of Real Textile Industrial Wastewater on Anatase under Solar Radiation

Hussein et al., (2011) have found that the rate of decolorization of Bismarck brown R on ZnO and different types of TiO_2 increases slightly with the increase of the temperature and the activation energy 24 ± 1 $kJ.mol^{-1}$ for ZnO and 14 ± 1, 16 ± 1, 21 ± 1 and 22 ± 1 $kJ.mol^{-1}$ for TiO_2 (Degussa P25), TiO_2 (Hombikat UV100), TiO_2 (Millennium PC105), and TiO_2 (Koronose 2073), respectively. Figure shows the impact of temperature on photodecolorization of Bismarck brown R by using TiO_2 (Hombikat UV100).

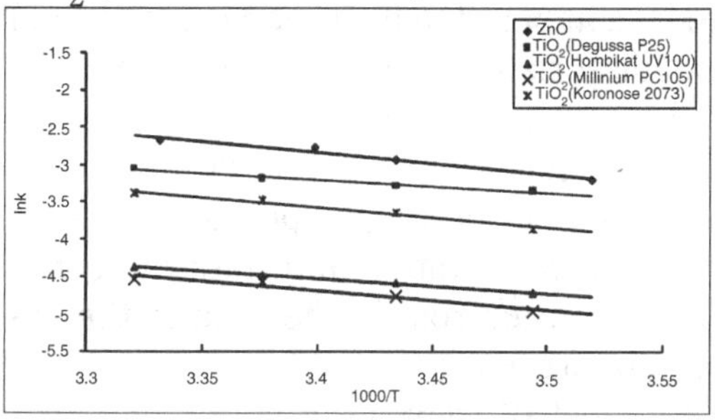

Fig. Arrhenius Plot by Different Types of Catalyst with Bismarck Brown R

EFFECT OF ADDITION OF OXIDANTS

It is well known that the addition of oxidants increases the rate of photocatalytic degradation of dyes by the formation of hydroxyl radicals (Salvador and Decker, 1984; Jenny and Pichat, 1991). However, this is not general for all types of dyes (Hachem et al., 2001).

Production of additional hydroxyl radicals occurs when hydrogen peroxide is added through the following mechanisms (Galvez, 2003 and Dong et al., 2010):

- Trapping of photogenerated electrons.

$$H_2O_2 + 2e^- \rightarrow 2OH^-$$

- Self-decomposition by photolysis

$$H_2O_2 + hv \rightarrow 2\dot{O}H$$

- Reaction with superoxide radical anion $O2^{\bullet -}$

$$H_2O_2 + O_2^{\cdot} \rightarrow \dot{O}H + OH^- + O_2$$

The addition of persulphate leads to form sulphate radical anion by trapping the photogenerated electrons (Konstantinou and Albanis, 2004)

$$S_2O_8^{2-} + e^- \rightarrow SO_4^{2-} + SO_4^{\cdot -}$$

The formed sulphate radical anion is a strong oxidant and reacts with organic molecules pollutants as follows (Galvez, 2003):

- Abstracting a hydrogen atom from saturated carbon.
- Adding hydrogen to unsaturated or aromatic carbon.
- Removing one electron from carboxylate anions and from certain neutral molecules.

Sulphate radical anion can also react with water molecule to produce hydroxyl radical (Konstantinou and Albanis, 2004):

$$SO_4^{-} + H_2O \rightarrow SO_4^{2-} + \dot{O}H + H^+$$

Other oxidants such as iodate and bromate can also increase the reaction rate because they are also electron scavengers, while chlorate has been proven insufficient to improve effectiveness (Galvez, 2003). However, these additives are too expensive to be compared to hydrogen peroxide and peroxydisulphate. Moreover, they do not dissociate into harmless products. The addition of oxidant to reaction mixture serves the rate of photocatalytic degradation by:

- Generation of additional •OH and other oxidizing species.
- Increasing the number of trapped photoelectrons.
- Increasing the oxidation rate of intermediate compounds.
- Replacement of oxygen role in the case of the absence of oxygen in the reaction mixture.

Table shows the effect of addition of hydrogen peroxide on the rate of photocatalytic degradation of red disperse dye on ZnO. The results indicate that the apparent rate constant increases with the increase in H_2O_2 concentration to a certain level and a further increase in H_2O_2 concentration leads to decrease in the degradation rate of the red disperse dye.

The presumed reason is that the addition of H_2O_2 to a certain level increases the production of hydroxyl radicals, but the additional amount leads to reduce the amounts of photoholes and hydroxyl radicals (Legrini et al., 1993; Malato, 1998; Daneshvar et al., 2003; Konstantinou and Albanis, 2004):

$$H_2O_{2+} 2h^+ \rightarrow O_2 + 2H^+$$

$$H_2O_2 + \dot{O}H \rightarrow H_2O + H\dot{O}_2$$

$$H\dot{O}_2 + \dot{O}H \rightarrow H_2O + O_2$$

Table. Effect of Addition of H_2O_2 on the Apparent Rate Constant of Photocatalytic Decolorization of Disperse Dye.

Conc. of H_2O_2 (mmol.L^{-1})	K_{app} (min^{-1})
0	0.0893
1.5	0.1408
3.0	0.1499
4.5	0.1680
6.0	0.1404
7.5	0.1290

This behaviour relates to the competition between the adsorption of organic pollutants and H_2O_2 on the surface of photocatalyst. The required amount of

H_2O_2 reaches the highest level of enhancement for the rate of photodegradation is related to the ratio of the concentration of organic pollutants and H_2O_2 (Galvez, 2003). When the pollutant concentration is low compared with the concentration of H_2O_2, the adsorption of organic pollutants decreases due to the increase of adsorption of hydrogen peroxide and, as a result, the additional hydroxyl radicals generated by H_2O_2 do not react efficiently.

COMPARISON BETWEEN MINERALIZATION AND PHOTOCATALYTIC DECOLORIZATION

Mineralization of dyes is a process in which dyes are converted completely into its inorganic chemical components (minerals), such as carbon dioxide, water and other species according to the structure of dye.

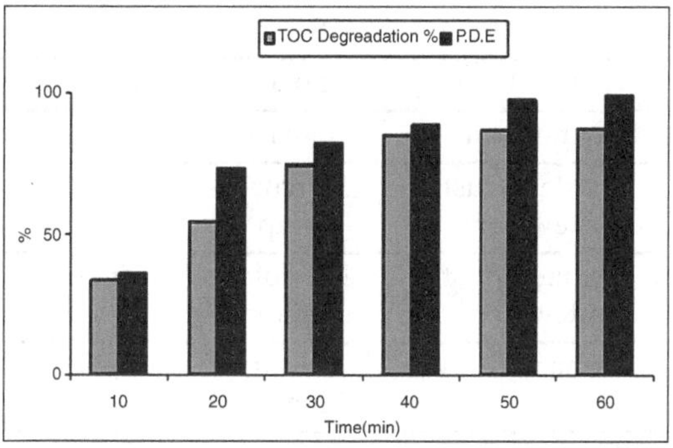

Fig. Comparison between Mineralization and Photocatalytic Decolorization of Bismarck Brown R on ZnO

Mineralization of Bismarck brown R was evaluated by analyzing total organic carbon (TOC) (Hussein et al. 2011). The results shown in Fig.15 indicate that photocatalytic decolorization of Bismarck R brown was faster than the decrease of TOC. The results indicate that per cent TOC reduction was about 73 per cent after 60 minutes of irradiation while the per cent of decolorization achieved 88 per cent for the same period of irradiation. These findings in good agreement with those reported before (He et al. 2007 and Chen 2009). This could be explained to the formation of some by product, which resist the photocatalytic degradation. Furthermore, the formed by products need more time to destruct.

EFFECT OF IRRADIATION SOURCES

Table summarized the obtained results from the different techniques used for the treatments of textile industrial wastewater and different types of industrial dyes (Hussein, 2010). The results show that the decolorization rate of textile industrial wastewater is faster with solar light than with UV light. The results indicated that solar energy could be effectively used for photocatalytic degradation of pollutants in wastewater.

Table. Effect of Irradiation Sources on Photocatalytic Decolorization of Textile Industrial Wastewater and Different Dyes

Process	Type of Treated Waste or Dye	Source of Irradiation	Type of Catalyst	Time for Complete Mineralization/ Hours
Photocatalytic	Textile industrial wastewater	Solar	ZnO	2.7
Photocatalytic	Textile industrial wastewater	Solar	TiO_2	4
Photolysis	Murexide	Solar	–	7.5
Photocatalytic	Murexide	Solar	ZnO	3.8
Photocatalytic	Murexide	Solar	TiO_2	3.3
Photolysis	Thymol blue	Solar	–	7
Photocatalytic	Thymol blue	Solar	TiO_2	2.7
Photocatalytic	Thymol blue	Solar	ZnO	3.3
Photocatalytic	Textile industrial wastewater	Mercury lamp	TiO_2	2.
Photocatalytic	Textile industrial wastewater	Mercury lamp	ZnO	3
Photocatalytic	Bismarck brown G	Mercury lamp	ZnO	1
Photocatalytic	Bismarck brown R	Mercury lamp	ZnO	0.8
Photocatalytic	Bismarck brown R	Mercury lamp	TiO_2	1.2
Photocatalytic	Textile industrial wastewater	Mercury lamp	TiO_2	3
Photocatalytic	Textile industrial wastewater	Mercury lamp	ZnO	1
Photocatalytic	Textile industrial wastewater	Solar	TiO_2	1.8
Photocatalytic	Textile industrial wastewater	Solar	ZnO	0.33

CONCLUSIONS

- Photocatalytic degradation techniques is the most efficient and clean technology.
- Textile industries have become worldwide. Thus, this method can be considered as a promising technique for providing formidable quantities of water especially for countries facing serious suffering from water shortage.

- The existence of catalyst and lights are essential for photocatalytic degradation of colored dyes.
- Photocatalytic degradation efficiency (PDE) of textile industrial wastewater is obviously affected by illumination time, pH, initial dye concentration and photocatalyst loading.
- Solar photocatalytic treatment has been proved to be an efficient technique for decolorization of industrial wastewater through a photocatalytic process and the transformation is practically complete in a reasonable irradiation time.
- In the countries where, intense sunlight is available throughout the year, solar energy could be effectively used for photocatalytic degradation of pollutants in industrial wastewater.
- The zero point charge is 6.4 and 9.0 for TiO_2 and ZnO, respectively above which the surface of photocatalyst is negatively charged by means of adsorbed hydroxyl ions; this favors the formation of hydroxyl radical, and as a result, the photocatalytic degradation of industrial wastewater increases due to inhibition of the photoholes and photoelectrons recombination.

5

The Control, Principles and Elements of Textile Design

Textiles are frequently made to be decorative and are used to embellish and decorate both people and objects. The designers responsible for such textiles have to balance many factors when answering a design brief. What is the fabric for? How must it perform? Who is the customer? What are the economic restraints? How is the fabric to be produced?, etc. Research into how people select products shows that colour and appearance are two of the most significant factors, with handle, performance and price coming lower down in terms of importance.

Textile designers therefore need to have a good understanding and sensitivity to colour and aesthetics. The same design elements and principles apply to design as to fine art, and an understanding of and an ability to use these are as vital to the designer as to the fine artist.

'Design is about relating elements, whether they are similar or contrasting and visually arranging an interesting unity with them. Shapes, forms, colours and texture all combine to become a unified whole which is commonly called a design.' Art and the act of creating are highly personal activities. In the first instance it may seem impossible to objectively analyse works of art.

In different cultures and at different times, styles of art and even the media used to express art have varied enormously. There are, however, certain basic design features that appear to be natural and common to all art.

DESIGN PRINCIPLES

In every design or work of art there are some or all of the design elements—varieties of line, positive and negative shape, three-dimensional form, occupied and unoccupied space, colour, value and texture.

The manner in which these elements are used and combined determines the quality of a piece of work. Thoughtfully balancing, moving, repeating, emphasising and contrasting the design elements can achieve a unified piece of artwork or design.

Balance

Balance is a sense of stability when applied to opposing visual attractions or forces. There is a natural desire for balance, and in nature balance is ever-present. In formal balance, the design elements are almost equally distributed. A design or composition that is divided in half so that one side is the mirror image of the other is said to have symmetrical balance.

In radial balance, the design elements radiate from a central point as the spokes of a wheel or the natural form of a daisy. Asymmetry uses informal balance; a centre line or point is ignored, with the design elements being balanced visually, rather than in a symmetrical manner. The position of any particular shape in a composition contributes to its strength.

A shape in the exact centre of a picture plane is at perfect equilibrium. Moving the same shape off-centre can increase or decrease its importance. In any textile design, each shape affects everything else. Exciting balance or imbalance is usually only achieved after frequent arranging and rearranging. The study of composition has been a fascination for many artists and designers. The Pointillist artist Seurat was as equally absorbed in the study of composition as in the study of light.

Movement

Movement and the portrayal of movement have always fascinated artists and designers. By careful arrangement of the design elements, the illusion of movement can be created. In optical art and designs the sensation of movement may deeply affect the viewer's responses. Some paintings can provoke dizziness by making it difficult for the eyes to focus on a central point.

Associated with movement is time. Pictures and patterns are capable of holding our attention for varying amounts of time. Some designs may be so subtle that these are barely noticed, if at all, by the viewer, while others can hold attention for much longer periods.

Repetition

Repetition occurs when elements that have something in common are repeated. When a design consists of shapes that are exactly alike, repeated in a uniform and regular manner, then that design tends to seem more formal.

By varying the shapes and the spaces between them, a more informal interest is created. The repetition of some of the elements within a design repeat can hold designs together. Repeated shapes make patterns. Many textile designs, because of the method of manufacture, will automatically repeat.

Emphasis/Contrast

Unity exists when all the elements in a design work together harmoniously. In a unified design, each element plays an equally important part.

THE PRINCIPLE OF DIGITAL PRINTING

In digital printing, small droplets of dyes, in aqueous solution, are sprayed on the fabric to reproduce the pattern. The viscosity of the solutions is low and therefore, to strictly keep the pattern profiles (this problem originates from the capillarity of textile material) and allow a good definition of the pattern, the fabric must be prepared with a special pad-wetting process with thickeners and auxiliaries (typical of traditional printing) and then dried. The printing operation will be followed by a steaming process (to fix the dye) and washing, like traditional printing.

Digital printing has several advantages if compared to traditional printing systems: no costs to produce screens, rollers or other printing equipment (production costs range approximately from 210 to 230-520 euro per colour), and the possibility of printing patterns in a few hours (not days or weeks) in all the desired colour variants. Furthermore, limited yardages do not excessively increase costs and patterns can always be slightly modified or customised.

For environmental protection and cost reasons, it is also worth considering that this system applies controlled quantities of dyes, thickeners and auxiliaries, avoiding printing paste residues (usually excessive when prepared with traditional printing methods) and their consequent disposal.

An inconvenience due to ink-jet printing is the low output efficiency and difficult reproducibility (by changing programme or printer), the reduction of the colour space and above all the poor penetration of the dye onto the fabric, with evident discrepancies between the two sides of the fabric. All these problems make this printing system suitable for specific uses.

DROP-SPRAYING TECHNIQUES

The Drop-on-Demand technique, *i.e.* the direct application of a drop of dye on the fabric is the most diffused digital printing technique applied to the textile field.

In particular, the nozzle technique used to spray the drop of dye determines the size of the drop itself, the spraying frequency, the accuracy, the evenness and partly also costs and reliability related to the machine. The type of nozzle used also influences the choice of dyes/inks used. *Thermal nozzle (HP Technology, Bubble-jet):* a small quantity of dye in aqueous solution is heated at 300-400°C inside a small container.

The steam bubble created causes the drop of ink to be forced through the nozzle and adhere to the fabric, while the quick cooling of the ink creates a concentration of dry ink particles and an immediate stop of the drop.

A very high jet frequency can be obtained with this system. This technique makes it possible to use plates with a very high number of nozzles at quite low cost but one of the problems lies in the application of special inks that can be used at high temperatures.

Furthermore the nozzles can be subjected to continuous and quite rapid deterioration due to deposits (originating from the decomposition of dyes and/ or precipitation of salts) produced by high temperatures in the steaming unit. We can briefly conclude by saying that this technique is scarcely reliable due to the colour variations connected with deterioration, and also because when a single nozzle does not efficiently work in a printing head featuring a very high number of nozzles (even though a single nozzle is quite inexpensive) the whole printing head must be replaced thus entailing process interruptions and higher costs.

Piezoelectric nozzle (Epson, Canon, Roland, Mimaki etc. technology): this technique is essentially that of ejecting drops of ink contained in a small unit by effect of the deformation of a crystal subjected to the action of an electric field.

This technique is more precise and reliable that the previous one because crystals are much more strong and hard-wearing than resistances and also because the system reliability remarkably improves thanks to the elimination of the deposit problem; furthermore less sophisticated (and therefore less expensive) dyes/inks can be applied with this system.

This technique makes it possible to vary the size of the dye/ink drop by varying the intensity of the electric field. It is still impossible to use plates (printing heads) featuring a very high number of nozzles, and each head is much more expensive than the previous ones.

THE PROCESS PRINTING TECHNIQUE

This system essentially reproduces design patterns by means of a technique similar to digital printing (dietering): by using four (four-colour printing with 4 printing heads) or seven colours (seven-colour printing with 7 printing heads) and by mixing and/or spreading on the fabric dye drops of different colours (this step is controlled completely by special software) different shades can be accurately reproduced. The intensity can be adjusted:

- VDS Process (Variable Dot Size by analogue modulation): the drop size determines the intensity of the colour shade.
- FDS Process (Fixed Dot Size by digital modulation): the number of drops per surface unit determines the intensity of the colour shade.

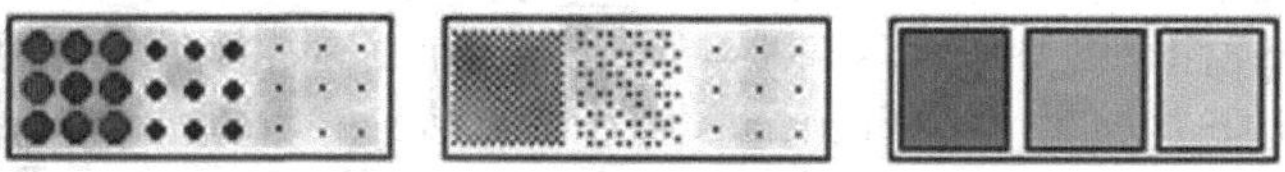

Process VDS *Process FDS* *Spot*

This system allows (by using 4 or 7 standard colours and the same number of plates with nozzles) a quite easy reproduction of a wide coloured area, but creates problems in reproducing patterns with backgrounds in pastel shades (marked dotting, poorly uniform backgrounds), and very unsatisfactory penetration of pastel colours (poorer quantity of liquid ink). Some problem could arise also for the combination of shades and light colours.

It is worth considering that the manufacturers of these printing systems are always more oriented towards nozzles granting more and more sharply defined printing (1200-2000 dots per inch), with always smaller droplets: this is an excellent technique for paper printing but not for textile printing, since very high definitions are not necessary and cause reliability problems (nozzles with very small holes which often clog), poor penetration and slow printing.

SPOT PRINTING TECHNIQUE

The colours to be reproduced are prepared by mixing master batches (similarly to traditional printing) and suitably diluted (a special colour kitchen must be set up). Therefore it would be necessary to use as many inks as the colours of the pattern and as many printing heads as the colours. This technique allows the printing of the fabric with the same number (maximum number) of ink drops per surface unit, by eliminating dotting problems, unevenness and enhancing the penetration in design patterns with background in pastel colours.

Obviously patterns with a huge number of colours cannot be reproduced (each colour must have its corresponding printing head and an excessive number of printing heads would make the printer less reliable). This makes the printing process very difficult and requires using a special colour kitchen, and washing the printing heads at every pattern varying, thus causing time losses and serious problems for nozzles, particularly for piezoelectric nozzles (drop ejecting holes with diameters of 10-12 micron).

DESIGN PRINCIPLES AND ELEMENTS

Design features, the elements and principles of design, can be called the language of art and design. Pick up one of the many books on design and these elements and principles will be there, although they may be detailed slightly differently in each. This is, in part, due to the way that in any one work of art no one element or principle can exist in isolation. Designing can be defined as relating and visually arranging components or elements to create effects. Space, line, shape, form, colour, value and texture are the design elements with which artists and designers work to create a design. The design principles of balance, movement, repetition, emphasis, contrast and unity are what artists and designers do with the design elements to make the art form or design.

Design Elements

Line

A line can be a mark made by a pen or drawing instrument or it can be any continuous mark that causes the eye to follow along its path. The viewer's eye travels along the path of a line because a line is longer than it is wide. A line moves, and as it does so it indicates direction. A straight line leads the eye swiftly across the picture plane but the eye travels more slowly when following the path of a jagged line. Lines appear in different ways. There are curved lines and straight lines.

These can be long or short, thick or thin, ragged, sharp, light, dark, simple or complex. Lines can be broken and yet have direction. Lines can be textured and can be coloured. Lines can be made from repeating similar elements, as diverse as dots or people, in a lengthways direction. Lines can be used to create form, to give depth. Lines can be carefully controlled to create optical sensations and can be used to project feelings of sensitivity and strength.

When someone looks at lines they try to fit them into something related to their previous experiences; for example, scribbled spirals may be interpreted as seashells, and a few lines can easily suggest an apple to someone who is familiar with this fruit. In some ways this can restrict the artist's or designer's freedom to make viewers see what is intended, although it does also open up new possibilities. Skilful artists and designers play with this tendency to see familiar forms in everything by using only a minimum of visual clues to evoke a far more elaborate response. Care must be taken, however, that abstract designs and patterns do not unintentionally suggest unpleasant or ugly forms to the viewer.

Shape

When a line turns and meets up with its start point, a shape is created. A shape or figure is a positive thing and occupies positive space. The area surrounding a shape is called the background or ground. It is a negative thing and occupies negative space. Shapes can be clearly defined with hard edges but often they are not clearly defined, which means that it is more difficult to see where shape ends and background begins. There are a great variety of shapes to be found in nature.

Many artists have drawn creative inspiration from natural shapes. Through their imagination, artists have invented new ways to use shape and communicate ideas. Shapes can be solid or opaque, linear, textured, coloured and outlined. Shapes can be transparent, revealing other shapes behind them. Similar shapes need not be identical, yet they can have a common relationship, which visually ties them together. Some shapes will command more attention than others, depending on their size, colour, value, texture, detail or their location in relation to other shapes. Tall shapes are elevating, long flat shapes express calmness, downwardpointing shapes activate the sense of falling.

Form

Form is structure. In art and design, form is the illusion of three-dimensional volume or mass seen in two dimensions. Careful observation of the forms around reveals that, in nature and in man-made objects, many forms can be described as combinations of the basic geometric structures; spheres, cylinders, cones, cubes and pyramids. Paul Cézanne, the impressionist artist, in his landscapes and still lifes, explored how these basic geometric shapes combined to make more complex shapes.

There are several methods that can be used to give an illusion of three-dimensional form and some of these are detailed below. Line, shape, colour, value and texture can all be used to suggest form by the artist on a picture plane, and by the textile designer within fabric.

Space

Space flows in, around and between forms or shapes. A flat surface has only two-dimensional space. This means that it has length and width but no depth. It is impossible to create actual depth or space on a flat surface but an illusion of space, distance or depth is possible. There are many methods used by artists and designers to create this sense of space, and to convince the observer that there is space and depth when, in fact, they are victims of a type of visual deception. Objects placed higher up can create the feeling of depth or distance.

Overlapping shapes can also create the feeling of depth, as can converging lines. If, in two-dimensional art and design, space exists purely as an idea or concept, it leaves the artist or designer free to compress or stretch it as they wish, to portray the particular feeling that is desired. Colours have an effect on space. Colours that are warm and bright appear closer, while dull or cool colours recede into the distance. A flat surface that is covered only with pattern can eliminate any feeling of space.

Colour

Under normal light conditions, well over 10 million different colours can be seen. Throughout history, the study of colour and the development of colour theories were frequently undertaken by artists and designers, and many artists have spent their lives attempting to understand colour.

Seurat wanted to apply a scientific system to the methods of the Impressionists, and his Pointillist theories, where dots of colour from a restricted colour palette were used to create the impression of a wide range of colour, clearly show his fascination with analysing colours. Colours communicate; some colours are associated with cold (blues and greys) and some with warmth (reds and oranges). Colour can convey the time of day, weather conditions and temperature, and even the time of year.

Colours can be designed to blend in with the environment or to stand out. Frank Lloyd Wright saw architecture as growing directly from the earth on which it stood and therefore used the colours of the surrounding areas in his buildings while Richard Rogers and Renzo Piano, the designers of Paris's Pompidou Centre, used colour, not to blend in, but to code the different pipe systems for heating and other services. In art and design work, the colour can be used in a natural or abstract way. Maps make use of both abstract and natural colour; natural colours are used when illustrating mountains, deserts and seas while maps showing political boundaries will use abstract colour.

Colour is a crucial part of all branches of design and design-based industries. Advertising agencies know that graphic information in colour will have a more profound effect than that in black and white. Colour grabs the attention span because memory recall from colour is quite pronounced. Colour is used to promote corporate identity. Blue is used by many banks to give the suggestion of reliability, while the appetite colours of red and yellow are used by fast food chains. Green is frequently used to denote environmental friendliness while more subtle greens convey a feeling of upmarket status. The *American Express* card was originally launched in yellow but was a flop; however, when its colour was changed to that of the American dollar bill it became a success.

The green livery of *Harrods*, the world-famous upmarket store in London, epitomises good taste and sophistication. Forecasting colour trends is itself an industry, and fashion colours change with the seasons. Underlying these cyclical trends, however, are some basic colour preferences. In colour popularity tests, blue is frequently placed first, with red second. Blue ties are most popular, with red ties in second place, and blue and red cars regularly occupy top positions, only being outdone by silver, a colour symbolising luxury and wealth.

The colour of a product influences the perception of it, and this is used extensively in marketing. Many people believe a red car will drive faster than a white one. Tests with coffee showed that coffee served in a red mug was preferred to the same coffee served in a yellow mug (considered too weak) and in a brown mug (too strong). Foodstuffs claiming to be pure and unadulterated often use blue and white packaging to communicate purity. However, in this age of environmental awareness it is predicted that more earthy colours will symbolise a natural product while white will be associated with chlorine and all that is environmentally unfriendly.

Products communicating strength adopt vibrant and contrasting colours; the greater the contrast, the stronger the associated power. Colour can be used to represent a product, with some colours idealising a product, and some using biological signals to communicate function. Yellow and black are used in nature as a warning symbol for reptiles and insects that have poisonous bites or stings and this colour combination is therefore often used to represent danger. Yellow and black is used on signs when the desire is to signal caution.

Colour Mixing

There are two ways to mix colour—'additive' and 'subtractive'. Additive colour is the mixture of coloured lights. The three primary colours of red, green and blue when mixed together in equal colours produce white light. Mixing the three additive primaries in differing amounts of coloured light can create any colour in the rainbow. Colour televisions use the principle of additive colour mixing. In subtractive colour mixing the principle is exactly the opposite.

The subtractive primary colours are cyan, yellow and magenta. When mixed together they subtract from the light producing black. When different pairs of the subtractive primaries are mixed, the colours red, green and blue are produced. These principles are used when we paint and in photography and colour printing.

Colour Systems

As the human eye can distinguish between 10 million colours, it is clear that to describe colour experiences by name is imprecise. While everyone knows what is meant by tomato red, different people have different ideas as to exactly what colours are meant by beige and sand; mauve and lilac. In order to accurately describe or pinpoint colours, a reference or colour system is required. All the attempts to notate colour can be traced back to the work of Sir Isaac Newton. In 1660 he re-created a spectrum by directing a narrow beam of white light through a prism; he went on to develop a colour wheel by taking the two ends and bending the spectrum into a circle.

This colour wheel evolved and changed over the centuries and in 1810 Otto Runge created a spherical model, with white at the North Pole and black at the South Pole and with Newton's colour circle forming its equator. In 1915 Williem Ostwald devised his double-cone colour solid; also in 1915, Albert Munsell developed another system of colour notation that added steps to the constituent hues. Munsell allowed his three-dimensional colour solid to respond in shape to the different potential strengths between hues, creating an asymmetrical colour solid. His colour model can be explained in three dimensions, in terms of hue, value and chroma. *Hue* is the 'colour' of a colour, *i.e.* its redness, greenness or yellowness, *value* refers to the amount of lightness or darkness, while *chroma* refers to its saturation or its colour strength. Anumber of commercial colour specification ranges have been developed from Munsell's work including the BS4800 and the Pantone and Colour Dimension systems. In textiles and clothing, the *Pantone* system is probably the most commonly used, and a whole range of products is available in its defined colours. These include ranges of marker pens, paper and even fabric samples.

Colour Psychology

There is a commonly held belief that red is exciting and green calming. But which red and which green? Research has shown that strong greens, yellows and blues are all seen to be as exciting as strong red. The same research also showed that pale and dark versions of the same hues have quite the reverse effect. Pink has a calming effect on violent people. A common belief is that blue increases the size of a room while red has the opposite effect. Recent research, however, has shown that it is not the colour itself that has this effect but rather its value, the amount of whiteness or blackness. However, with regard to temperature, colour does have a very definite effect. Red and yellow rooms are perceived to be warmer than blue and green rooms.

Value (Tone)

Between the whitest white and the blackest black there are countless degrees of light and dark values. Value is important to the designer. Shapes that are close in value or tone appear to merge together. Visually dark values appear to come forward and light values tend to recede, but the reverse can occur. Sharply contrasting values attract attention and the use of light against dark or dark against light can create the illusion of size difference. Value is probably the most elusive of all the design elements. The success or failure of a piece of work may rely on the use of values within it.

Texture

One definition of texture is the surface of a material as perceived by touch. Running a hand over a surface may find it to be smooth, rough, dull, glossy, hairy, sandy or bumpy. Texture is very important in textiles, and how fabric feels is an important consideration in product choice. Texture can be seen as well as touched and artists and designers can use a variety of techniques to convey textures when none is actually there. Repetition of design elements can often create a visual illusion of texture.

GRAPHIC OR MULTIMEDIA DESIGN

Graphic design is the knowledge of typography, composition/layout, print mediums, packaging, branding and identity, web design and motion graphics. A graphic designer produces corporate and brand identity, web site design, motion graphics, multimedia, package design and prints. As a result of the rapidly changing computer technology, multimedia design is becoming a design occupation on its own. Multimedia designers produce video/film, animations and audio for commercial as well as educational use.

INDUSTRIAL OR PRODUCT DESIGN

Industrial design is the professional service of creating and developing concepts and specifications that optimize the function, value and appearance of products and systems for the mutual benefit of both user and manufacturer. Industrial designers maintain a practical concern for technical processes and requirements for manufacture; marketing opportunities and economic constraints; and distribution sales and servicing processes.

They work to ensure that design recommendations use materials and technology effectively, and comply with all legal and regulatory requirements. Some industrial designers are calling themselves "product design engineers", emphasizing the technological aspect of their background. In addition to supplying concepts for products and systems, industrial designers are often retained for consultation on a variety of problems that have to do with a client's image.

Such assignments include product and organization identity systems, development of communication systems, interior space planning and exhibit design, advertising devices and packaging and other related services. Their expertise is sought in a wide variety of administrative arenas to assist in developing industrial standards, regulatory guidelines and quality control procedures to improve manufacturing operations and products. Industrial designers are generalists, not specialists, in contrast to the other design professions.

FASHION DESIGN

Fashion design is applied in the garments industry as well in haute couture. As the name implies, fashion design is about designing apparel and fashion accessories. A fashion designer knows fashion illustration, pattern drafting, design, draping, clothing construction, textiles, fashion history and production techniques.

TEXTILE DESIGN

Textile design is the knowledge of (natural and synthetic) fibres and their properties, textile constructions including woven, knitted and non-woven, how colour and designs are applied, finishes and testing for quality and performance. A textile designer is good at sourcing and developing new fabrics for products, but is often specialising in designing woven or printed fabrics, carpets, towels, bed linen or knits.

JUTE

In Bangladesh, 25 per cent of the population (30-35 million people) is directly affected by jute production and many more by the industry. Jute manufacture is one of the most important industries, as important as ready-made garments, seafood processing and pharmaceuticals. India is the world's largest producer of jute and also its leading consumer of jute, selling 80 per cent of the production on the home market. Bangladesh is the only other major producer of jute, exporting 80 per cent of the yearly output. Bangladeshi jute is supposed to hold a high quality. Jute has, however, a low price image in the local market.

Jute is a natural fibre obtained as an extract from the bark of the jute plant. It can be harvested three times a year and provides a necessary break between rice crops helping to enrich the soil. The Bangladesh Jute Research Institute claims that jute can be a substitute for cotton.

Bangladeshi agriculture has as its foremost aim to provide food for the rapidly growing population. Rice is the main crop. In fact, Bangladesh hardly produces cotton and silk for its textile production, but has remained a net importer of yarn at least since 1947. Almost all yarns are imported; warp silk from China, cotton from China and India. If jute, which is locally grown, could be used for more of the textiles being produced in Bangladesh, this would be of advantage both to the farmers and to textile producers like jute mills, garment manufacturers, handloom weavers or craft businesses.

The jute industry is, unfortunately, a so-called "sunset" industry. Decline in sales and production volume has been a fact for many years now. Synthetic substitutes play a great role in the downfall of the jute industry, as many of the traditional jute products, *i.e.* packaging and carpets, now have been substituted by synthetics. The government policy of subsidizing the government-owned mills, has given the private-owned mills a difficult competitive situation while not avoiding a major breakdown of the government-owned production.

JUTE DIVERSIFICATION

Parallel to the sun setting over the jute industry, significant research and development (R and D) activities have taken place in many countries in the world during the last twenty years. As a result, many technologies and processes are now available for commercialization and more are coming out. The jute sector has previously concentrated its efforts on producing packaging (sacks and Hessian).

The diversification efforts have put an emphasis on:

1. Home textiles
2. Agro textiles
3. Composites.

The Bangladesh Jute Research Institute (BJRI) claims to have developed 37 new technologies, of which 15 are ready for commercialization on a pilot scale, among them jute blankets and prayer mats, knitwear, jute denim and blended jute/cotton thread. The BJRI is concentrating its efforts on research and leaves the commercialization to the private sector.

An example of jute diversification in the field of composites in Bangladesh is a student project at the Department of Materials and Metallurgical Engineering at Bangladesh University of Engineering and Technology (BUET). The aspiring engineers are experimenting with jute as a substitute for glass in "glass fibre", hoping to provide a useful material for (at least) the furniture producer who is lending the technology to the university. Realising there is a need to build bridges between the R & D environment and the jute industry, the Ministry of Jute with financial support from the European Commission, set up the Jute Diversification Promotion Centre (JDPC) in 2002.

Perhaps the most well-known diversification programme was the Jute Development Programme in India launched in 1992 by UNDP in order to search for and develop diversified, high value added jute-based products and then to encourage entrepreneurs to commercialise results through an incentive credit facility. This project managed to achieve production of blended yarns where jute was mixed with cotton, linen, ramie, viscose and polyester in union blends. Critical voices, say, however, that the UNDP programme failed because *"They helped the jute mills in design and development, but other aspects like investment in new machinery, changing attitude of workforce, marketing, market study, investing in new technology (and) machinery, etc. was missing from the project."*

DIVERSIFICATION IS NOT NECESSARILY COMMERCIALIZATION

Several of the private-owned mills in Bangladesh have already to some extent diversified their production. The main challenge at present does not seem to be diversification as such, but commercializing the results of the already existing research from the R & D environments. The process of commercialization is a complex weave of trial and error; analyzing market potential, designing products, finding technology, financing investments etc.

CO-OPERATION BETWEEN R&D INSTITUTIONS AND THE INDUSTRY

The contact between R & D and the businesses seem to be on an ad hoc basis, taking place when the businesses face technological questions they hope the research institutions can solve. The mills, as well as smaller companies, are rather critical towards the government-financed research institutions, claiming inefficiency and little willingness to meet the needs of the businesses.

JUTE AND THE SMALL BUSINESSES

Small businesses working with jute are mostly found in the crafts sector. This sector is large; it is supposed to be the second largest source of employment in the rural areas, even if the businesses are small. Most often the businesses are organised with a small number of employees (for instance 30-40) and a large number of connected craftspeople (for instance 300-400) from which the products are bought. In most developing countries and certainly in Bangladesh, the concept of the craftsperson as a "crafts-artist" is not known. The Bangladeshi craftspeople are production workers, not artists.

National Crafts Council of Bangladesh (NCCB) expresses a strong interest in jute as a material in crafts production. For the handloom businesses the use of blended threads in the production of fabric for both apparel and interiors are interesting. One business has already had success with fabric and apparel of jute/cotton mixes and claims that jute is a selling point. For crafts businesses in general both blended fabrics and pure jute, treated or not, are materials that can be utilized for product development. New products for export markets are especially demanded. This is an area where NCCB believes it is necessary to use international designers, hoping that donors will be willing to subsidize the cost.

MARKETS

For the smaller businesses, the local market is the most important one. It would be of great advantage for the smaller businesses if the jute mills succeeded in selling and producing larger volumes of, for instance, the soft fabric. The price would go down, making the fabric available for many more of the small businesses.

Then the fabric (and eventually also softer yarn) could be used for products not only aimed at the export markets, but also at the local and regional markets, giving income to a larger group of companies and, incidentally, work to more people. However, some high value niche export markets are approachable also with the present high prices, as has been proved in the Norad project, if the quality is high enough.

THE NORAD PROJECT "THE DEVELOPMENT OF THE JUTE INDUSTRY IN BANGLADESH"

The NORAD jute project started in 1997 and ended in June 2003. In the end, the participants were 2 mills, Sonali and Pubali, and four craft producers: two craft businesses, Esheeta and Swajan Crafts, and two NGOes, Concern and CORR - The Jute Works. The project has succeeded in producing a new, soft jute fabric in light and heavy weights and in different colours, lately also with prints. The market focus has been "home furnishings". The fabric and the home furnishing products have been presented every year, starting 2001, at the Heimtextil fair in Frankfurt, Germany. A greater quality consciousness has grown out of working with niche products for an upper-grade market.

The Dutch design company LA Colours worked as consultants in the project's first years. The project participants grant much of the honour for their success to the Dutch designers. The companies participating in the Norad project have now joined in the loosely knit "Jute Group". South Asia Enterprise Development Facility (SEDF) has been asked by Norad to make a market study, especially looking into the potential of the soft fabric that was a result of the Norad project.

THE JUTE MILLS

Government owned mills dominate the jute sector, both in terms of employment, number of mills and capacity. The mills are large industrial plants, often employing thousands of people, most often using old technology.

In the 1950-ies and 60-ies, mainly four types of jute industries were set up:

1. Hessian and Sacking Cloth and Bags (spinning and weaving machinery)
2. Spinners (spinning machinery)
3. Carpet Backing Cloth (spinning and weaving machinery)
4. Jute Carpet Mills

According to E. Sigvaldsen et al´s report (2002), there are 47 spinning mills, 11 privately owned weaving mills and 30 government-owned weaving mill. The government does not compete in the spinning segment.

STRATEGY OF VALUE-ADDING

The focus of the jute mills presently seems to be on value-adding, of finding ways to add value to the products that can be made with (more or less) the existing technology. The mills are today more interested in finding new markets for thread and weave, and in diversifying – and adding value - within the framework of their existing production of thread and weave than starting on a radical diversification process involving great economic risk.

Yarn and weave represent the major part of the existing jute industry and possess an unused potential for product development. The large markets of agro- and geo textiles can be approached using (mainly) existing technology. The soft fabric will have to be further developed so that problems with shrinkage and colour fastness are solved. Thinner threads and blended threads are other areas of growth.

The idea of developing fully vertically integrated manufacturing, e.g production of thread and yarn, fabrics, cutting and sewing is done in one place, has been taken up by some of the Bangladeshi jute producers. In a longer perspective, at least some of the mills show interest in expanding beyond the traditional thread and weave, for instance are non-woven textiles mentioned as an interesting opportunity.

LARGE VOLUME MARKETS

The jute mills need volume sales in order to keep their businesses going and will have to focus on markets accordingly. New products should (1) utilize large volumes of jute, (2) have a sustainable demand for large volume output in the foreseeable future and (3) be profitable.

As pointed out by Vinay Chand in his report about the Norad jute project, the textile markets approachable for the treated fabric are large, - much larger than the specialised interior design markets where the products from the project's small businesses and NGOes can be sold. The jute mills are themselves pointing at large potential markets, expressing interest in producing geotextiles and agrotextiles.

DESIGN IN BANGLADESH

"Why scramble to be the cheapest when using design can make you the best by transforming everything you do?"

DESIGN COUNCIL (UK)

The strong position of the garment industry in Bangladeshi economy - it is one of the largest export industries - does unfortunately not indicate that the country in general has a lot of excellent fashion or textile designers. Buyers from developed countries usually bring their own designs as well as their own materials to the Bangladeshi garments industry. The buyers focus on price and quality, not on contributing to create a thriving design profession in Bangladesh or elsewhere. In the jute industry, which is another important source of export income, one of the informants said outright that the employees can copy, but not design. In general, the manufacturing industry of Bangladesh seems to be very good at copying, but lacking in the understanding of how their competitiveness can be increased by using design as a strategic tool in reaching existing and new markets.

The main characteristics of the design situation in Bangladesh seems therefore to be a great capacity of copying as well as being able to produce cheap products that are designed elsewhere. Even if this "strategy" of copying and producing cheaply seems to function at the present level, it is not a strategy that will help Bangladeshi exports in the future. The role of Bangladesh will be reduced to the role of the re-active, the one player in the game who is not trying to win, but just - barely – trying to participate. Such an approach is understandable for a poor country, but is unfortunately doomed in the long run. If you are not winning markets shares, you are loosing them.

INSPIRATION FOR TEXTILE DESIGNS

Inspiration for textile designs can come from a variety of sources. It is possible to create textile design work by going straight to the process being used. For example, a weave designer may go straight to the weaving shed, choose warp yarns in colours that appeal and, with a knowledge of fabric structures, make an attractive fabric. Experience, however, has shown that such a method of designing often results in fairly mundane design. Innovative, exciting textile design starts with much more fundamental paperwork; drawings and paintings exploring colours, textures, shapes and patterns.

Drawing from objects in an imaginative and open way will inspire new colour combinations, textural ideas, shapes and arrangements of these. Inspiration for such paperwork, and ultimately for textile designs, can come from many things. Both natural and man-made objects can inspire. Whether the source material is natural or man-made, abstract or tangible, how themes and source material are used is a matter for debate. Drawing from man-made items such as architecture can raise the question as to whether or not the resulting design work is truly original. Drawing from architectural patterns can be seen, if the work is very literal, as merely copying someone else's designs — the proportions and shapes have been created by another person.

Working from nature, however, is usually seen as producing more original work. The designer may still be copying but is constantly making decisions and judgements about what to include and what to omit. Being influenced by the work of artists and other designers raises the question— when does a design influenced this way become a copy? Is a textile design that lifts elements from other designers or fine artists good or bad design? Fine art can inspire textiles; part of a menswear collection by the fashion designer Paul Smith in the late eighties was made in fabrics that were inspired by the paintings of Matisse.

Textiles can be created by fine artists; tapestries and silk scarves have been designed by artists and sculptors such as Picasso and Henry Moore. Textiles can have fine art directly applied; reproductions of the Mona Lisa have been applied to a variety of products, from tea towels to tee shirts. Much of textile design is derivative. Source books for designers often consist of designs from the past or from other cultures. William Morris, the famous nineteenthcentury writer and designer, often based his designs on Persian textile designs and many of the designs credited to him that are still best-sellers today were almost literal copies of the original Persian ones.

Does this knowledge change how Morris's work is perceived? Does using archive or historical sources allow copying to become acceptable? Whatever the inspiration or source for a textile design, the textile designer needs to have a good understanding of colour and aesthetics. A textile designer must be able to draw. They must understand and be able to use colour and pattern.

PATTERN

Any textile designer needs to have a good idea of how their design is going to look as a finished fabric. One piece of paperwork could be worked through to become many different finished designs, woven, knitted or printed. All designers need to understand how designs can repeat. While many designs work as straight repeats, where the design element is repeated again and again, one on top of another, side by side, there are also half-drop repeats and tile repeats. Mirroring, where elements are reflected about either, or both, x and y axes, is another common way of repeating elements.

Fabrics may have sections within one complete repeat that show other repeat patterns, while some fabrics such as border designs will employ repeated designs only within certain sections of the fabric. It may be useful to mask-off certain areas of paperwork (or sample blankets and sample fabrics in the case of woven and knitted fabrics respectively), or use a frame to establish the most suitable sections to develop.

It is important to consider the repeat structure from an early stage in the design development, and various sketch plans of possible repeat structures should be evolved. Very often, trying to interpret paperwork too literally can be a problem, particularly for a print designer. By its very nature, weaving and knitting design requires elements to be extracted from the initial inspirational paperwork.

For print designers, this decisionmaking process of what to include and what to discard is equally important. The best designs are so often simple. This advice is as true for printed textile design as for any other type of design. Everything that does not contribute, everything that is not necessary, should be left out.

THE ACTIVITIES OF A TEXTILE DESIGNER

The overall purpose of a textile designer is to design and produce to an agreed timetable, an agreed number of commercially viable textile designs. Research into what designers do has identified several different activities.

The activities listed below relate to an in-house fabric designer engaged on branded, contract and general work:

- Deciding what to design.
- Producing original design ideas.
- Developing design ideas through to a form suitable for initial sampling.
- Supervising the production of original fabric samples and keeping suitable records.
- Submitting samples to customers, to company or to company selection systems.
- Adapting and modifying designs from sketches or fabrics submitted by customers (or obtained from within the firm or group) to meet a price or other restriction.
- Obtaining the acceptance of a certain proportion of the original sample fabrics by the customer.
- Controlling the production of sample ranges in consultation with the customer's technicians, technologists and buyers.
- Establishing production specifications.
- Ensuring the production of sample ranges by certain agreed dates.
- Reporting to the company on contacts with customers, competitors, exhibitions, developments by fibre and yarn producers, etc., in order to augment market intelligence.
- Controlling the systems and procedures for the storage and retrieval of designs and samples within the company.
- Working within an agreed departmental budget.
- Controlling the supply of materials and equipment for design and sampling.

All textile designers will certainly be involved in some of these and should have an understanding of all of them.

HOW DESIGN WORK IS DONE

Design ideas are developed through in different ways, depending upon how an individual designer likes to work, upon the product being designed and upon the manufacturing process/es. From initial paperwork, designers of constructed textiles may take their design ideas and develop them through on the loom or knitting machine.

Designers of printed textiles will usually develop their ideas right through on paper. 'Much design work is carried out in a very direct and informal way.

The degree of formality becomes a function of scale and the number of interests represented:'

- Design is an investigative process; it involves research. The first stage in any design exercise is normally an enquiry into what the client (or potential client) requires; their needs and expectations.
- Design is a creative process; it involves art and aesthetics. Designs can be copied or invented. Adesign problem is solved with the help of know-how, ingenuity, pattern recognition abilities, lateral thinking, brainstorming, etc.
- Design is a rational process; it involves logical reasoning in the checking and testing of proposed solutions, information analysis, experimentation, field trials, etc.
- Design is a decision-making process; it involves making value judgements.

The selection of particular combinations and configurations, layouts or shapes involves considerable uncertainties. These are resolved by estimating the values that are likely to be placed on the various major alternatives. In this respect the task of the designer is closer to that of a manager rather than a researcher, scientist or artist. Estimating the social consequences of various design alternatives also falls into this category. Solving design problems involves a mixture of the intuitive (having ideas) and the systematic (rational scientific appraisal). There is no single solution to a design problem, rather there are many solutions although some may be better than others.

The Design Process

The design process usually starts with a requirement or desire for a new item or product. Research will usually be carried out then to find out as much as possible about this need and about the role or function the new item or product is to have. Ideas generation is the next stage when various alternative initial ideas are conceived.

These initial ideas are then usually developed through until the designer is happy to offer them as proposals to meet the initial need. In the early stages, alternative ideas will often also be presented. These proposals will be considered and perhaps modified. A decision is then taken as to the best solution to the design problem and the necessary specifications and instructions will then be given.

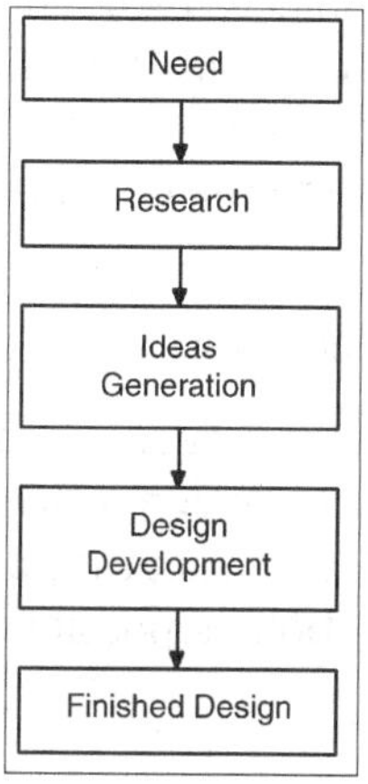

Fig. The Design Process.

Planning Design Work

Any design project will usually need to be completed by a certain date, and more often than not designers will find themselves working on more than one project at any one time. Efficient project management and efficient time management are obviously both desirable. A project entails one or a number of individuals working together over a period of time to achieve an agreed goal or outcome. Any project needs to be completed on time, to a standard and within a budget. A design project, as any other project, needs to be planned and managed efficiently.

The Planning Process

Planning can be described as managing and controlling events to achieve a goal and making the best use of resources. Planning is achieved by making decisions about what is to be achieved (aims) and how best to get there (who does what, when and how). While planning takes a little time, lack of planning can result in the waste of a lot of time, the expenditure of much money and the generation of considerable stress and tension. A plan gives what could be described as a set of route markers that allow checks to be made as work is undertaken to achieve the goals. Regular checking of current position against plan identifies any deviations from course which can then be remedied. The more detailed the plan, the more route markers there will be. A plan helps in the controlling of events rather than events determining direction. Very few jobs go exactly according to plan, but if there is no plan there is no way to keep control and track progress.

Objectives

An objective describes something to be achieved in the future. The essence of an objective is that it defines what is wanted without describing how this is going to be achieved. Without objectives people do not know where to invest their time or resources, and they can drift in an unco-ordinated way. When objectives are unclear, it is impossible to make effective plans and people will not be committed to action. Despite the obvious importance of objectives many people find it difficult to set them.

Identifying the Aims and Objectives of a Design Project

The aims and objectives of any design project are usually identified at a briefing meeting. This is when the designer finds out what is required of them. Many clients often have little idea of what they actually do want so it is up to the designer to get as much information as possible; this is usually done by asking questions.

Checklists

Using a checklist can be very helpful at many stages in the design process. For example, a checklist for a briefing meeting would simply be a list of questions, prepared beforehand, that the designer will use as a prompt. Enough knowledge of a situation is required to enable a checklist to be developed that covers the right questions. Use of a properly prepared checklist should ensure that the designer does not forget to ask for any necessary information and this should help avoid misunderstandings that might create problems later in the design process.

Project Planning Methods

Different methods can be used to help plan design projects. The simplest is to arrange elements of the plan in a logical sequence. Working from the brief, designers determine what is required to fulfil this and plan a logical sequence of events to take them through the design process to a design solution that satisfies the brief.

Backwards Planning

Often, the best way to ensure that design work achieves the requirements of the brief is to start by examining the expected final products or outcomes. These products or outcomes need to be listed and agreed. Having considered what it is that is to be achieved, the processes or work to achieve the outcomes can be determined. Finally, it must be decided what inputs are necessary to enable the work to be carried out successfully.

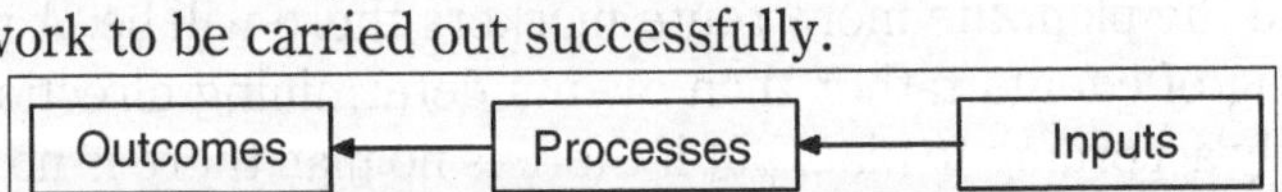

Fig. Backwards Planning

- *Outcomes*: The planned result of the project, equal to the aims and objectives of the project, to achieve the quality or standard required.
- *Processes*: The work needed to be done to achieve the output deliverables or outcomes.
- *Inputs*: The resources in terms of individuals and their skills, time, materials, equipment, techniques, etc., necessary to achieve the outcomes.

For example, a designer is asked to design a range of printed curtaining fabrics. The *outcomes* would be a specified number of fabric designs, with colourways, for a specific time, to sell at a specific price point. The *processes* would be initial research of the market (to see what is currently selling and what the competition are doing), the formulation of initial design ideas, the development of these ideas, the selection and production of the specified number of fabric designs (with recommendations for print base fabrics) to be included in the range, colourway development and selection, and the specification of final fabrics for production.

The *inputs* would be the designer with the required skills and knowhow, the time they would require and the materials and equipment necessary.

Gantt Charts

A slightly more sophisticated way of planning is to use a Gantt chart. During a design project, several stages can be under way at the same time. A Gantt chart is a simple horizontal bar chart that graphically displays the time relationships of the stages in a project. Each step is represented by a line or block placed on the chart in the time period in which it is to be undertaken.

When completed, a Gantt chart shows the flow of activities in a sequence, as well as those that can be under way at the same time. Gantt charts can also be used to chart actual progress, by drawing lines in different colours to show the start and end dates of each step. This allows easy assessment of whether or not a project is on schedule.

Network Analysis

This is a generic term used for several project planning methods, of which the best known are PERT (Programme Evaluation and Review Technique) and CPA (Critical Path Analysis).

These are more sophisticated forms of planning than Gantt charts and are appropriate for projects with many interactive steps.

Project Management

Managing a project involves co-ordinating activities so that they run according to plan. The progress of a project should be monitored and measured against the plan. When deviations occur, corrective action should be taken.

Time Management

All projects are time-bounded and for any practising designer it is desirable (if not essential!) to make the best use of the time available. Time management is simply making the best use of time to achieve what is necessary. To effectively manage time, goals and time limits need to be set. What is required? What has to be achieved and when? How efficiently do these goals have to be met? It is only against set targets that success can be measured. The way any individual uses time is unique to that individual. Some people use time as chunks into which they can fit certain activities, all neatly stacked. Others have no clear view of time, selecting activities at random or changing priorities to suit the current crisis. There are strengths and weaknesses in both ways and it is of value to consider both because many people will alternate between the two, depending on the jobs in hand.

The way designers use their time will be different, since everyone has their own pace of work and their own rhythms, with different peaks for different activities. Where possible, work should be done at times to suit an individual's own speed and their own way of organising and completing activities. Managing time costs time. To sit down and plan the best use of time is an investment.

It takes time to learn to use software packages for word processing and database management but their use is a huge investment for future time management. The time spent on any activities should be considered afterwards and evaluated. Can any lessons be learnt to plan more effective use of time in the future? *The Pareto principle: 20 per cent of what you do yields 80 per cent of the results.* Targets that will result in a high pay-off should be identified. Constructive avoidance is when time is spent on work that is neither important nor urgent in preference to urgent and important work.

Time management involves:

- *Planning tasks*: Getting information, assembling relevant facts, skills, experience, resources, establishing what is known or what needs to be known, processing the information, considering options and identifying the risks involved,
- *Stating what has to be done:* Scheduling what will be done, how, where, when and by whom.
- *Getting Things done:* Doing things and monitoring progress against checkpoints/standards, reviewing outcomes by assessing the results achieved in relation to the aims, determining if more needs to be done, analysing successes and difficulties so as to plan for improvement.

RANGE PLANNING

A range is a group of products offered for sale. These products will be the designer's answer to a brief. Having the right products in a range is very important in terms of how well that range will sell.

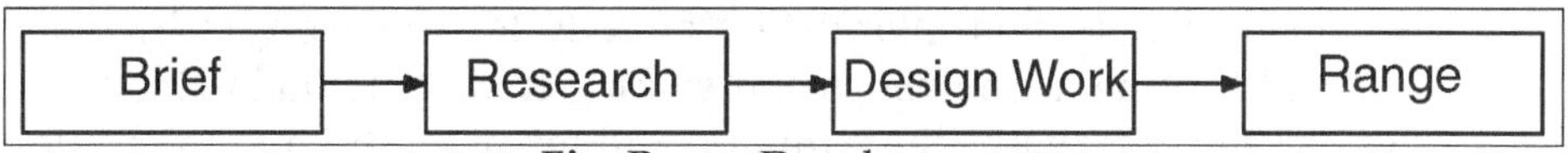

Fig. Range Development

Good design adds value to a product. With products of equal quality and price, the design will be what differentiates. What does the designer need to know or make decisions about when planning a range? Or, put at its most simplistic, how does a designer decide what to design? All designers are designing for the future and are influenced by trends in their product area.

The whole area of forecasting is one where much money is invested to 'get it right' and much money can be lost if a designer gets it wrong. As well as needing to understand forecasting, a designer needs to know other information when planning a range. This includes:

- Type of products,
- Customer/market areas,
- Price points,
- Manufacturing capabilities,
- Quality required,
- Number of designs,
- Number of colourways.

It is usually the job of the design manager (or stylist) in a studio to plan the ranges. There will have been some discussion with sales, marketing and production about all or some of the factors listed above—essentially to build up a full picture of what criteria the range must answer. The type of designs depends on the products being designed for, the customer profile, price points for sales, manufacturing plant availability and quality.

The number of designs and colourways in a range is controlled by several commercial factors including the amount of capital the company wishes to tie up in the range, the raw materials that are needed, the amount of stock that can be carried, and the production capacity, *i.e.* the number of machines available for printing/knitting/weaving the range (some machines may be tied up with specific customer orders rather than general range production).

Research

After the briefing meeting, the next stage in the design process is usually research: finding information about products and markets, production processes and techniques, etc.

Information can be collected in different ways:

- *By waiting to see if information presents itself (gathered by chance from contacts and perhaps by reading)*: This is unsatisfactory if there is any form of time limit on when the answer is required.
- *By asking questions*: This is appropriate if it is known who to ask and what to ask. Known as primary research or field research, this is

where designers find information directly for themselves; by going out and looking at the shops and visiting trade fairs and exhibitions, by asking relevant individuals and groups, by using mailed questionnaires, by setting up focus groups and by interviewing people.
- *By looking up what is required*: This is appropriate if it is known where to look, how to look it up and what exactly is being looked for. Known as secondary or desk research, this is where information is gleaned from work already carried out by other individuals or groups. Such research might include looking at market research information obtained through government censuses and large market research programmes.

Information Search

To ensure that the required information is found and time is not wasted, the collection of information should be systematic. The aim should be to find key information. Identify what is not known.

As far as possible, vague requests for 'information on' some problem area should be broken down into questions defining what it is that is not known and what needs to be found out.

The questions that need to be answered should be identified and listed. The level of answer required should be established (there is a balance between the importance of what is being looked for and the time and effort spent looking) and a checklist of likely sources of reliable answers to questions can be prepared.

It is sensible to keep control over any search for information by setting limits and deadlines and by using expert opinions and expert literature searchers to identify the most promising sources. It is important to stop as soon as there is enough information to move forward and it is essential to keep accurate references.

Getting Information from other Organizations

There are many occasions when designers need to find out information from other organisations. Sometimes, the information sought may have to be bought, sometimes it may be free—part of a marketing strategy on behalf of the supplier.

However, whether visiting a new supplier to source yarn or buttons, or telephoning to check prices of fabrics, it is important that designers know exactly what they want so as not to waste time. Before any visit, the information required should be clearly identified.

It is essential for any designer to be prepared, and establishing a checklist of questions that need to be answered is a good way of making sure that all relevant information is obtained.

When seeking information about market areas and products, it would be usual to employ professional market researchers. However, sometimes designers may prefer to carry out some market research themselves. In either case, well-constructed questionnaires can be very useful, and when market research is being carried out by professionals the designer's input should not be overlooked.

Questionnaires

Questionnaires should be designed in such a way that they take a minimum amount of time to fill in. A busy company director or buyer is much more likely to complete and return a questionnaire if it is clear to read, to the point and does not take up too much of their valuable time.

Formulating questions in such a way that they can be answered by ticking boxes is much more likely to elicit a positive response than having questions that require written replies.

Ideas Generation

With some initial research undertaken, the designers need to be able to generate suitable ideas to help them answer the brief.

Design ideas can come from almost any source, but sometimes designers may find difficulty in thinking of suitable sources and themes for work.

Brainstorming

Brainstorming is a very useful method of generating ideas and can be used at many stages in the design process. In its most formal sense, it is a group participation technique for generating a wide range of ideas in order to tackle a stated problem.

In a less formalised way, an individual or pair can use a brainstorming session to generate ideas. What follows is a description of brainstorming at its most formalised; such a rigid approach, however, may well inhibit ideas:

- *A problem statement is formulated*: Too vague or too restrictive a statement should be avoided.
- The group of people to participate in the session is selected. The group should ideally include some people familiar with the problem area. Small groups of about 4–8 people are best.
- Five or ten minutes are allowed for group members to write down their first ideas in reply to the problem statement.
- Group members are encouraged to continue writing down new ideas whilst each person in turn reads out one idea from their set. The session rules are:
  - No criticism is allowed of any idea.
  - Crazy ideas are welcome.
  - The more ideas the better.
  - Ideas should be combined and built upon.
- After the session, the ideas are evaluated.

THE DIVERSITY OF TEXTILE DESIGN AND TEXTILE DESIGNERS

The diversity of the textile and clothing industries is reflected in the many different types of designers needed. At every stage of manufacture of textiles there are colourists determining the fashion colours in which the fibres will be produced, yarn designers developing yarns to meet certain requirements, knitted-fabric designers, woven-fabric designers, carpet designers, print designers, embroidery designers, knitwear designers, designers of women's wear, men's wear and children's wear, accessory designers, designers of casualwear, sportswear, eveningwear, swimwear and designers for the mass market, haute couture and designer labels, etc.

The Purpose of the Textile Designer

The role of the designer can be quite complex but the overall purpose can be stated quite simply —the textile designer has to design and produce, to an agreed timetable, an agreed number of commercially viable fabric designs.

Depending on the markets that he or she is designing for, several different activities are involved in fabric design and the number and type in which any designer is involved will vary according to the product and production methods used, and the type of company for which the work is done.

Stylists

Designers also put together ranges. For example a stylist might handle the development of a company's range of printed fabrics. A range is a group of fabrics (or products) designed, developed and edited to be shown and sold to the market each season.

The stylist initiates the design work, organises and directs the development and colouration of intended designs (frequently using freelance designers), and co-ordinates with manufacturing personnel to have samples made.

These samples are shown to customers; the stylist then edits and finalises the group of designs that will form that season's range. Further down the chain, buyers and merchandisers in retail organisations do much the same range-building processes.

Colourists

Some designers work purely with colour, predicting colour trends and putting together palettes of colours for specific seasons and product groups.

Other colourists will work further down the design process line, colouring designs produced by other designers to create different and alternative colourways.

Repeat Artists

Acompany producing printed textiles will often employ designers whose main function is to take designs and put these into a size and repeat appropriate to the intended enduse.

In-house and Freelance Designers

Designers may work for manufacturing companies as in-house designers, or they may work independently as freelance designers. In-house designers, or as they are sometimes called staff designers, are employed by a company usually on a full-time basis, although some may be employed part-time. Often they work within a manufacturing environment, although they can also be employed by retailers and by converters. Freelance designers may either work for independent studios or through an agent, producing designs on paper for which the studio/agent receives a commission when the designs are sold to mills and converters. Alternatively, freelance designers may put together a portfolio of their designs, which they may sell directly to stylists.

The work they produce for their portfolio, while it will have at least to reflect trends, will be often very much what they themselves like and want to produce. While the designer may have a view of the type of customer who will buy their designs when the design is developed, there may be no specific customer waiting to buy their work on completion. Freelance designers may also develop design work according to a stylist's specification.

For example, a freelance print designer who is particularly good at intricate florals may well be approached by a stylist to work on a specific print idea that will form part of that company's new season's collection. The brief may include size details, colouring details and even the type of flowers to be painted. This work will be commissioned in advance. The designer develops their paperwork with the knowledge that when it is finished there is a buyer for it.

Freelance weave and knit designers will normally work on a specific project with a manufacturer. They will be commissioned to produce a range of fabrics, or, in the case of a knitwear designer, a range of knitted-garment designs. A third type of designer found within textiles is the consultant designer. A consultant is employed by a company to advise on design matters and may be given the task of managing the design programme.

A consultant designer will usually work for several companies at any one time, although their contract may be such as to impose restrictions on their working for closely related organisations.

Very often consultants will do little actual working-through of design ideas themselves; rather they will make design policy decisions and direct other designers who may be in-house or freelance. All designer systems have advantages and disadvantages to the designers themselves and to the organisations for whom they work.

Just as many different people doing different jobs are given the name designer, so too are many rooms called design studios. The 'studio' can be anything from an area set aside on the factory floor to a large, smart office or even suite of offices. One designer or a whole team of designers can work in a studio and a few or many of the functions of, and stages in, the design process might be carried out there.

TIMING IN THE TEXTILE AND CLOTHING INDUSTRIES

Products are planned and produced well ahead of retail selling seasons in all areas of the textile and clothing industries. In August, and even July, autumn clothing appears in the shops. Before this clothing can be designed, fabrics must be designed and shown to clothing manufacturers, and orders placed, produced and delivered. Before then, new yarns and sometimes even new fibres must be developed. The scheduling and amount of time necessary for all of these steps are dependent on the amount of change occurring in the product, the volume being produced and the efficiency of the producing companies. At a minimum, textiles are usually designed a year to a year-and-a-half in front of the retail season. Major changes such as the development of a completely new type of fabric will take even longer.

Seasonal Ranges

Manufacturers producing for the fashion industry will normally launch two main ranges a year; for the selling seasons Spring/Summer and Autumn/Winter. Sometimes, midseason ranges are also produced. Manufacturers of furnishing fabrics and household textiles normally produce one main range per year. In every segment of the market, the higher-volume manufacturers work further ahead of the retail season than the companies producing lower volumes (at the higher end of the market). Generally, smaller operations do not require the long lead-times that are necessary for large-volume production runs.

PRINTED AND CONSTRUCTED TEXTILES

Textile designers may be categorised by the types of product or fabrics for which they design. Printed textiles are often considered to include fabrics patterned by dyeing techniques as well as those where the design is applied to the fabric by a printing process. Constructed textiles include woven textiles, knitted textiles, lace and carpets.

From Sketchbook to Fabric Samples

The design process is about the realisation of ideas —ideas that are transformed into tangible products. The designer will start with some sort of brief describing the project. For the in-house designer, the briefing may be very informal as their experience within the company will usually mean that they will already have an understanding of the requirements of the range.

They may, indeed, be the person setting the brief! Hopefully, any brief, formally or informally delivered, will have all the information needed, but if not, the designer must be ready to ask questions. Who is the customer? What is the price range? What are the manufacturing possibilities in terms of machine availability? The designer needs to get as much information as possible. Designs come from a variety of different sources, and different designers will have different methods for developing design work from initial ideas.

Most designers keep some form of sketchbook —something they carry with them and use to record and store ideas to call upon as a source of inspiration in the future. This may contain drawings and photographs, magazine cuttings and fabric pieces. Often, a brainstorming session will be used to give inspiration and generate a wide variety of ideas. Some of the considered themes and ideas will be selected and subsequent artwork based around these. While the way these ideas are used will be individual to the designer, much of the process is similar for many different types of designers.

Yarn Design

Yarn designers have to know what machinery they have available to them and what that machinery is capable of producing. They will usually start with ideas collected on their travels; colour ideas, texture ideas, looks and moods that they see as becoming important in the future. They should have an understanding of their existing and potential customers and what they want. The development of a range of yarns may involve the creation of a completely new yarn using a newly-developed fibre. It may involve the tweaking and fine-tuning of existing production ranges to meet the requirements of a future market.

An existing yarn quality may require to be re-coloured so that it can be used to service a fashion market. Some ranges of yarns will run from season to season servicing a classic market area. For example, a range of knitwear yarn for a manufacturer of schoolwear will include the classic school colours — navy, maroon, grey, bottle green, bright red and bright blue.

The knowledge base that the yarn designer will have built-up is the springboard from which they work, in conjunction with technicians and development personnel, looking at the overall range, determining qualities and the types of yarns appropriate to their market area. The selling cost of the yarn must also be considered. Colour is very important and yarn designers will often subscribe to colour prediction publications to help them determine their colour palettes. The same colours or shades of colours will often run through the different yarn types. Co-ordination is as important in yarn design as in any other area, helping the customer to see how the product can be used and often prompting more sales.

Yarns may be designed to work with other qualities in the range. Sample yarns will be specified and made using different components and in different colourways until the designer is happy with the handle and appearance. The yarns may have to perform in specific situations and they will have to be tested to ensure that any such performance requirements are met.

Weave Design

After the briefing meeting where the work that is required is established, and after initial ideas have been collected, one of the first things that a weave designer will do is to establish a colour palette. This colour palette may change as design work develops but such changes are normally minimal. A colour palette will usually consist of groups of colours, chosen with regard to trends and predicted directions. The way these colours work together will be important. Adegree of flexibility is usually desirable as economic constraints will normally demand that stocks of yarns (both for sampling and production) are kept to a minimum. Once an initial colour palette has been decided upon, suitable yarns are then selected.

These will normally be selected from ranges offered by yarn producers and frequently from the colours in which the yarn producers have decided to run their ranges; sometimes, however, colours will have to be specially dyed. From their initial paperwork, weave designers will usually take their design ideas and develop them on the loom. Sample warps will be made up and different weave, colour and yarn combinations tried. Shape and pattern are important but these might well evolve more with the weaves used.

The way woven designs repeat is determined by the way the yarns are threaded up and the way these threads are lifted. This is a very integral part of the weave design process, and a good understanding of fabric structures is vital to allow the weave designer to be as creative as possible within the constraints set by the type of loom that is being used and its patterning capabilities. Sample blankets, or 'section' blankets as they are also known, consist of different sections across a warp. These sections may vary in colour, yarn type and/or weave. Initial sample warps will often have several warp sections with different drafts, and different warp patterns and yarns.

Different lifting plans (which, along with the way the loom is threaded, determine the weave) will be tried out as will a variety of weft patterns and yarns, again in sections. As well as true designs where the planned warp will be woven with the intended weft, there will be many areas where crossings occur.

These are sections where warp and weft patterns combine more by chance than by design. Areas of interest will be developed and hopefully an initial sample blanket will provide plenty of ideas for subsequent development. Having considered initial ideas, further sample blankets will be developed; these will show full repeats of the proposed fabric designs and will often have sections in alternative colourways. Some weave designers will work on computer-aided design (CAD) systems, using these to try out ideas before the weaving stage.

Many of the sophisticated weave CAD systems currently available, as well as showing representations of what the fabric will look like, also have facilities to show proposed weave designs in situ, on garments and on other products.

However, in practice it is still usual to weave actual fabric samples of those ideas which best meet the brief. The way fabric feels, its handle, is still a very important factor in fabric selection.

With a selection of ideas chosen, the making particulars or fabric specifications will be recorded so that these designs may be produced again. The yarns and colours used, the draft (the way the loom is threaded), weave, lifting plan (the way the warp threads are raised), warping and picking plans (colour and yarn arrangements in the warp and weft directions) all need to be recorded, as do any subsequent process such as applying a fire-retardant finish or raising. Throughout the weave design process, the designer is constantly having to make decisions, on colour, yarn, weave, size of repeat, etc. These decisions are made from a knowledge base that includes intended selling price, customer preferences, the range as a whole, co-ordination requirements, performance requirements, and so on. Once the design ideas are developed through into fabrics, these will usually then be presented in-house to the sales team, and then to customers.

Knit Fabric and Knitwear Design

Again, as in weave design, the design of a new season's range will usually start with the development of a colour palette. This will come from initial ideas generated in response to information from a briefing meeting. Suitable yarns will then be selected. The working through of design ideas is very much restricted by the type of knitting machine being used and the type of structures that can be created on it. Inspiration from drawn work will influence the textures, colours and proportions of a knitted fabric. Creating visual representations of sketchbook work is only possible when machinery with jacquard patterning capabilities is used.

Some designers will put their initial design ideas straight into graph form, ready for translation by a knitting mechanic. Others may do more paperwork, perhaps developing fabric and garment styling ideas side by side. The fabric ideas worked out, the designer will then usually decide which yarns should be tried in which fabrics. It is usual to knit several swatches to enable the designer to see how any fabric looks in different yarn and colour combinations. For the knitwear designer, concerned with the design of final garment as well as the fabrics, styling ideas must then be worked out.

The designer of cut and sewn knitwear will probably sketch out styling ideas, and the next stage will be to cut a pattern. A first sample will be cut out and made up, pressed, measured and tried for fit. Any necessary alterations will be made and a second sample made to check that these alterations and the final specifications are correct.

To help knitted fabric and knitwear designers, there are specific CAD systems available which allow representations of fabrics to be viewed before knitting. They also often allow fabric ideas to be mapped onto garments, and these facilities should cut down on actual sampling time. In practice, this is not always the case as the ease of changing things on the computer often results in requests for more initial work giving more choice. Once the garments and colour ways that will form the range are decided upon, the range is then ready for presentation to sales teams and customers.

Print Design

The print design process can take many different forms. With modern technology, sketchbook work can be taken and re-created on fabric very much as it appeared originally. Print designers, after initial research, will usually work their ideas through on paper. Any medium can be used in the artwork for print design, but materials with which designers can achieve facsimiles of particular desired effects in manufactured fabrics allow design problems to be more accurately resolved in the studio before the design is put into production. Print designs are sometimes worked in repeat from the start but often they are designed in balance and put in repeat later.

A balanced design is called a 'croquis' (a French word meaning *sketch*). A croquis should give the impression that would be seen if a frame were placed over any section of the finished cloth. Although not in repeat, a croquis will give the feeling of being in repeat. The most common repeat structures are simple block repeats, half drops and bricks. Print designs for apparel fabrics are usually designed in croquis form with the repeat sizes varying enormously. Upholstery and curtain fabrics have standard repeat sizes so these are often designed as a repeat pattern with the size being carefully considered from the beginning.

Often, the involvement of a print designer will end with the production of the design on paper, the preparation of colour separations and the approval of these usually being undertaken by the printer. However, in-house print designers will usually be responsible for approving the strike-offs (sample prints on fabric). Again many print designers (particularly those working for large multinational companies and corporations) will often find themselves working with CAD systems.

Such print systems allow designers to develop and expand design ideas readily. Design ideas can be scanned in, and repeat, cut-and-paste and other functions allow ideas to be manipulated quickly. Many systems also have colour separation facilities to aid in the preparation of screens. Some systems even allow sample fabrics to be produced, and this is a very valuable tool, making it much easier for print designers to take their ideas right through to fabric.

Design Adaptation and Modification

As well as producing ranges, textile designers will have to adapt and modify existing designs, design ideas and sketches submitted by customers. Here, textile designers are using their creative and technical skills to translate ideas into reality. They must, of course, be careful not to copy or adapt designs that have copyright, or are registered, without the express permission of the copyright holder or design originator.

Examples of Textile Design Briefs/Problems

An in-house designer working for a manufacturer of printed fabrics may be asked by a customer selling babywear to develop a small design with chickens as the motif. The babywear company may also employ a designer and they might have put together rough sketches and colour ideas to give an idea of what is wanted. In this case the brief will be fairly specific. A company may have seen a range of designs for duvet covers that they like. However, they may be unhappy with the colourways being offered.

An in-house designer working for the printer may be asked to re-colour the designs, producing colourways that are considered more suitable. Perhaps these will have to co-ordinate with a range of lampshades being developed in a specific range of plain shades. A knitwear manufacturer might find themselves approached by a company which has a sweater that they are manufacturing overseas and are currently importing.

The company is, however, experiencing real problems with quality and late deliveries. Essentially the company wants the knitwear manufacturer to re-create their sample sweater to a specific price point. The designer's job in this case would be to source suitable yarns and develop specifications for the production of a garment that is essentially the same as the original sweater in terms of quality and appearance but which also meets the pricing requirements.

A company might have a fabric that is currently selling very well — their 'bestseller'. The marketing director and his team require a 'new'version of this fabric, perhaps more up-to-date and with a co-ordinate. They are happy with the existing colourways but are open to suggestions regarding colour and scale. As their market area is becoming increasingly competitive they would also like to see the new version having elements of 'added value' if at all possible. This might mean repositioning the product in the marketplace or developing a new product that acts as a co-ordinate with the current best seller. In this case the textile designer's job is to develop a fabric using the existing colourway that fulfils the company requirements.

6

Ironing

INTRODUCTION

Ironing is the work of using a heated tool, or tools, (an iron) to remove wrinkles from fabric. The heating is commonly done to a temperature of 100°Celsius. Ironing works by loosening the bonds between the long-chain polymer molecules in the fibres of the material.

While the molecules are hot, the fibres are straightened by the weight of the iron, and they hold their new shape as they cool. Some fabrics, such as cotton, require the addition of water to loosen the intermolecular bonds.

Many modern fabrics (developed in or after the mid-twentieth century) are advertised as needing little or no ironing. Permanent press clothing was developed to reduce the ironing necessary by combining wrinkle-resistant polyester with cotton.

Ironing may also be used as a germ/parasite killing hygienic operation, such as in areas where the Tumbu fly is prevalent.

While nobody knows precisely when the practice began, people have been smoothing clothes with a variety of items, including glass, wood, and stones, since before recorded history. The first known use of metal to "iron" clothes, however, is known to have occurred in China.

The electric iron was invented in 1882, by Howard Seeley. Seeley patented his "electric flatiron" on June 6, 1882.

PURPOSE

Some people consider ironed clothes to be more aesthetically appealing than wrinkled clothes[*who?*].

After stitching or sewing new clothes, especially pants and suits, tailors iron them to give the clothes the appropriate shape. Fabrics such as linen are considered to be more comfortable following ironing.

Ironing will also kill vermin such as body lice, scabies mites and Tumbu fly, although the same thing could be accomplished more efficiently by machine drying.

RECOMMENDED IRONING TEMPERATURES

Textile	Temperature	Dot mark
Toile	240 °C	
Triacetate	200 °C	
Cotton	204 °C	* * *
Linen		* * *
Viscose/Rayon	190 °C	* *
Wool	148 °C	* *
Polyester	148 °C	*
Silk	148 °C	*
Sympatex		*
Acetate	143 °C	*
Acrylic	135 °C	
Lycra/Spandex	135 °C	
Nylon	135 °C	

Dot mark	Temperature
*	< 110 °C
* *	< 150 °C
* * *	< 200 °C

Lower temperature than above may be advised in some cases where the colour might be sensitive.

IRONING BOARDS

Ironing boards don't have to be hidden away in a closet or cloistered in the laundry room. As mundane as it may seem, there's a lot to learn about selecting a quality ironing board for great sewing and pressing results.

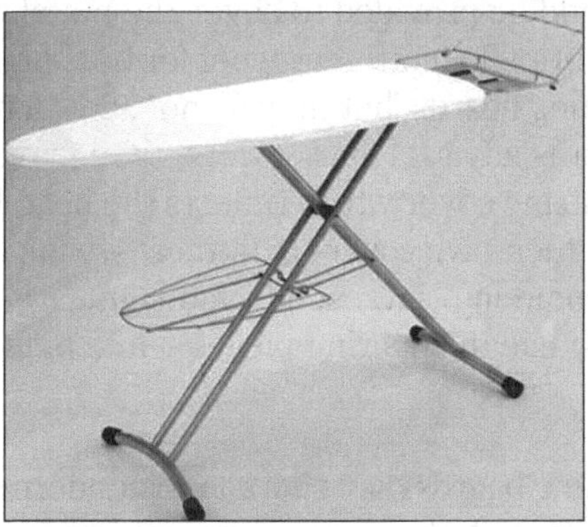

BOARD BASICS

Ironing boards can be freestanding or mounted, depending on the available space and frequency of use. Look for over the door models that fold down for use, wall-mounted fold-ups, hide-away boards protected in their own slim cabinets, tabletop versions, portable take-to-class types and the traditional two-legged versions.

TAKING SHAPE

Traditional ironing boards are long and skinny, about 15" wide and 4 1/2' long. They offer a narrower end for pressing hard-to reach areas. Sizes vary depending on the brand and some companies offer much larger and wider boards for pressing quilt tops and home décor projects. Others offer board toppers to retro-fit a traditional size ironing board to increase the surface area. Ironing board heights should be adjustable to avoid back strain. Most offer multiple heights than snap into position for stability.

TERRIFIC TOPS

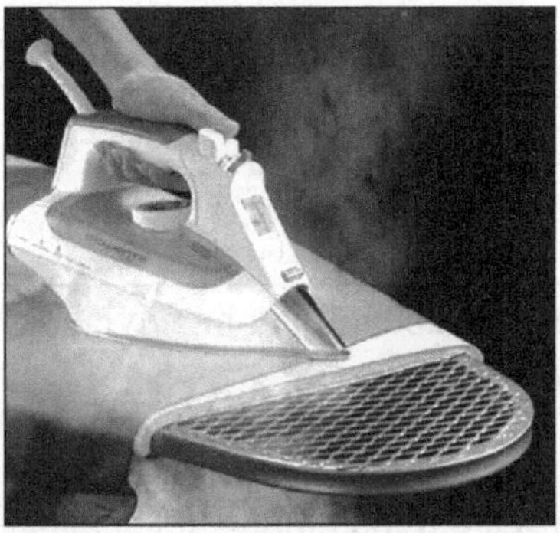

A good ironing board should have a smooth, flat metal surface with vent or grid holes to allow steam to escape and not be trapped in the project being pressed. Padding should be provided to cover the metal and it should be made from a natural fibre fabric like cotton or wool for best pressing. Foam pads are available for most ironing boards, but they do not allow for moisture absorption. A cotton or wool cover is advisable for sewers, though many boards come with a Teflon-coated drawstring cover which reflects the heat. Because of this, many sewers prefer to tailor their own covers to fit their sewing room ironing surfaces. Ready-made covers come in prints, solids and some are printed with grids for sewers and quilters to use in squaring pieces while pressing.

ADD-ONS

Good quality ironing boards may sport a rack underneath the board surface for storing laundry, press cloths or pressing equipment, etc. Others offer an attached iron rest at the end of the board, so the iron can be taken off the actual pressing surface and stored when not in use.

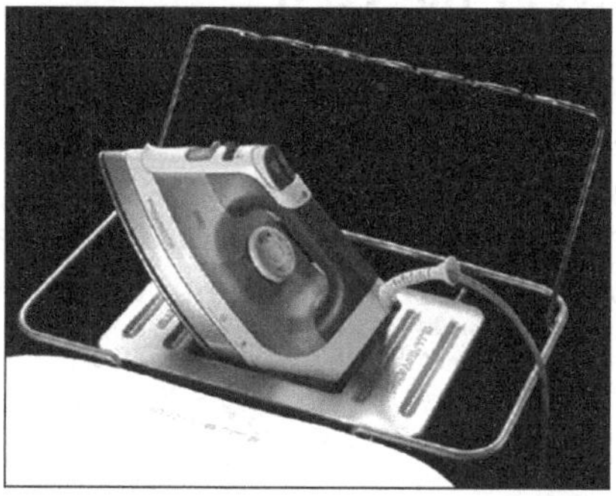

A garment rack may be incorporated at the end of the ironing surface for hanging freshly pressed items before they make it to the closet or to a sewing client. Iron cords can get in the way and some ironing boards have a cord holder which elevates the cable to keep it off the pressing surface. These usually screw on to the underside of the board, so can be removed when pressing extra-large projects. Non-skid feet are a handy feature if the ironing board lives on a hard-surface floor.

VACUUMTABLES

When is an ironing board also a vacuum? It's not a silly joke, but a reality that sewers love. Vacuum ironing tables use a fan to pull steam down through the surface (and the project being pressed) for quick cooling and skilled shaping on tailored projects. The suction also helps to hold slippery fabrics in place.

Most vacuum tables also have a reversible motor so that air can be pumped upward through the project. This is a helpful feature for pressing delicate areas like puffed sleeves, fabric that can be easily damaged like velvet or embroidery, and also to raise nap on fabrics like velvet and corduroy so that the iron doesn't leave lasting impressions on the pile surface. Coupled with steam generator ironing systems, these boards are the ultimate in pressing perfection and maximum steaming for tailoring and other sewing tasks requiring moisture and quick cooling. Most vacuum boards have wheels for easy mobility and they fold up for storage.

PORTABLE PADS

If pressing on the go is something you do often, check out the myriad portable pressing pads that fold up for travel.Whether you're taking them to class or on a family vacation, these handy surfaces substitute for a full-blown ironing board. Some come with gridded surfaces for being sure your bias is straight or other tasks where measuring is key, and still others flip to a rotary cutting mat on the reverse side.

4 REASONS TO LOVE IRONING

1. WELL PRESSED =WELL DRESSED

A well-ironed shirt means you care about the details. Details make a difference. The details get you noticed. With the best tools available, you take pride in doing things yourself – like preparing a memorable meal or creating the perfect garden. Ironing your own clothes shows that same pride. Remember, *how* you wear your clothes is just as important as what you wear.

2. CLOTHES HAVE FEELINGS TOO

Well, maybe not feelings, but they'll definitely let you know when they've been mistreated. Sending your cherished clothes to most dry-cleaners often gives you average results: shine, shrunken collars and broken buttons. Doing it yourself at home with the new style, pressurized steam irons and specialty ironing tables will give you professional quality results. Whether it's your favourite gingham shirt or that custom silk blouse from a trip abroad, your clothes will look better and last longer, without the added cost

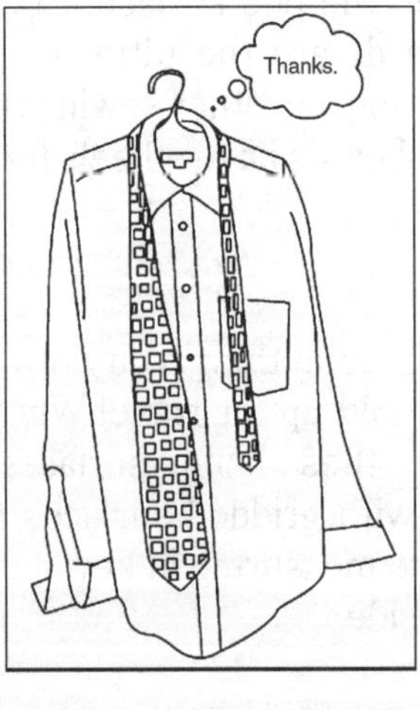

3. THREE SAVES

Laundering and ironing clothes at home won't save the planet but it will reduce the use of energy and exposure to harmful chemicals like Perclorethylene – the traditional agent used in dry cleaning, now banned in many areas. Most of your clothes can be laundered at home with excellent results using an energy-efficient front-load washing machine, then finished with a pressurized steam iron and vacuum/blowing ironing table. For garments that must be 'dry' cleaned, we recommend one of the new 'wet' cleaning processes that use 100 per cent water instead of chemicals like "Perc." It costs a bit more, but aren't your family and the environment worth it? Save money. Save energy. And help save the planet.

4. ZEN AND THE ART OF IRONING

The kids are asleep. Your home phone, cell phone and hand-held devices are turned off: Peace and quiet. You may not want to talk to anyone else today, but you do want to get something done in this perfect state of calm. Enter ironing. Many people swear by its meditative qualities: the repetitive movement of the iron; the hiss of steam; the disappearance of wrinkles – one by one. If that's too quiet, listen to some great music while you treat your clothes to a steam – generated by a premium Reliable™ iron – or turn on the TV for company. Before you know it, a pile of wrinkled laundry turns into a crisp, clean wardrobe for the week ahead. And you… are as refreshed as your clothes.

STEAM IRON

A steam iron uses superheated water to eliminate wrinkles in clothes and fabrics which may not be suitable for traditional dry ironing. Distilled water is usually poured into a holding tank and special heating elements convert it to steam. This hot mist comes out through a number of holes in the soleplate (bottom heating element) of the steam iron. As the steam loosens the individual fibres of the clothes, the steam iron's pressing action smooths out wrinkles.

Many traditional dry irons contain a steam iron option. Water is poured into a small reservoir and superheated until it becomes usable steam. A mechanical switch on the handle of the iron allows the user to select the steam setting. The steam itself should come out of small holes located near the tip of the pressing plate. For most conventional ironing needs, this combination of dry and steam iron settings should suffice. The steam generated by a traditional dry iron may not be overwhelming, but it will loosen wrinkled fibres.

Some ironing jobs such as curtains or quilts require a much more generous supply of steam. This is where a true steam iron can be useful. Distilled water is placed in a larger reservoir in a steam iron. Users can push a button to receive a burst of steam when needed. More holes in the soleplate means a more generous supply of steam while ironing heavier materials like suits or curtains.

Advanced steam iron systems may also feature handheld wands which can direct clouds of steam in and around vertically-hung garments or curtains. The steam itself should eliminate wrinkles without the need for a pressing element. Consumer rating boards suggest that the ability to generate 'vertical steam' is a major plus for a steam iron system. Some dry iron/steam iron combinations can only generate steam when the iron is held horizontally over an ironing board. Good vertical steaming means the steam iron can be used to iron curtains and quilts in place. Another important feature to look for in a steam iron is a means to control the level of steam. Some models only allow for an on/off situation—either the user has steam or not. Better steam iron systems have a dialing mechanism which can adjust the intensity of the steam produced. Consistency of steam is also a consideration. Once the water has become superheated, the steam should be steady and have some power behind it. Effective steaming means penetrating individual fibres, not merely wetting the surface.

IRONING-MACHINE

The ironing-machine, or, as it is often called, the mangle, is another device for making laundry work easier. It may successfully take the place of the hand iron for a larger part of the family ironing. There are two types of ironing-machines on the market:

1. Cold-roll ironing-machines, in which the rollers between which the garment passes are made of wood and are unheated, depending on their weight and pressure to remove wrinkles;

2. Hot-roll ironing-machines, in which one roll is cold and is covered with a blanket and cloth, just as for an ironing-board, and the other roll or concave plate is made of smooth iron and is heated.

The cold roll revolves against the heated metal plate. This is the more economical and satisfactory ironing-machine, although its original cost is greater. The hot-roll machine may be heated by gas, or gasoline by a slight change involving a small expense. If power is available, the machine may be run by power. The use of one of these machines reduces greatly the time required to iron in the usual way. Garments with gathers and sleeves cannot be thus ironed to look perfectly smooth and well shaped, but all bed and table linen, towels, handkerchiefs, stockings, such underwear as may not require perfect smoothness, kitchen aprons and the like, may be done successfully and satisfactorily.

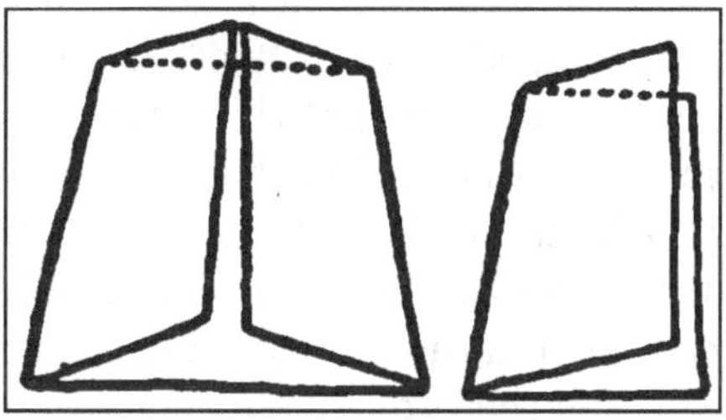

Fig. Another Method of Folding Sheets

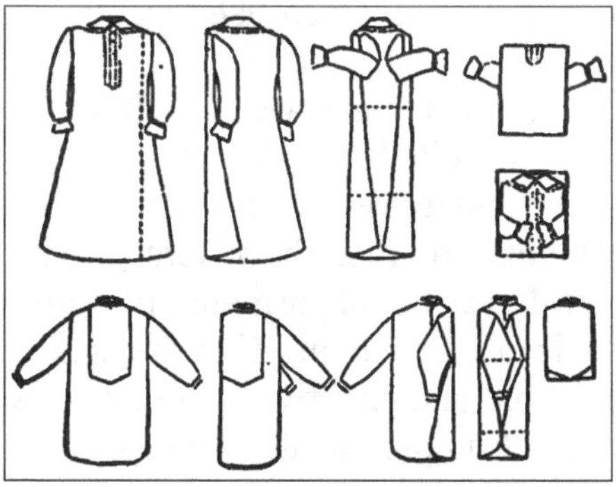

Fig. A Method of Folding Nightdresses and Shirts

7

Alternative Clothes Cleaning

WET CLEANING

Wet cleaning is both an old method and a new method for cleaning garments. Water has been used to clean clothes for centuries. Inventors, scientists and engineers have also been working for centuries to discover new methods to clean clothes quickly and efficiently. By the 1930's, the dry cleaning method became very popular as an affordable and efficient method for cleaning a wide variety of garments. One concern with dry cleaning, however, is that it uses chemicals that can be hazardous to garment care workers and to the environment. Because of these hazards, scientists and engineers began looking for a cleaning method that does not use hazardous chemicals. They came up with a number of new methods which are addressed in Module 5. One new method is machine wet cleaning, which uses water and detergents to clean clothes.

It is not the use of water that makes wet cleaning a new method for cleaning clothes. New computer-controlled wet cleaning machines, new dryers, new detergents and new spot removers are what make wet cleaning a new method. The new washing machines and dryers have controls that allow them to safely and efficiently clean a wide variety of garments in water. New detergents and spot removers are made of ingredients that are safer for workers and the environment, yet are as safe and effective at removing soils, stains and odors as dry cleaning solvents. Equipment, detergents and skill all contribute to successful wet cleaning

In this course, participants will learn about cleaning garments using the the wet cleaning process. It is hoped that after taking this course, participants will be motivated to add wet cleaning methods to their daily operations.

WHY DO WE NEED A NEW WAY TO CLEAN?

The professional garment care industry is currently experiencing one of the greatest periods of change since it began. Two forces are driving this change: technology and regulations. The garment care industry is working on new equipment, new detergents and new cleaning methods. Some of this new technology is available now. Some of it is still being developed and tested. Environmental regulations are forcing many industries to change their processes. These regulations have also affected dry cleaners.

Dry cleaning using organic solvents such as camphene, benzene, gasoline and Stoddard solvent began in the mid-1800's. Since then, the garment cleaning industry has continually changed to keep up with consumer demand, new developments, and new regulations. The 1930's and 1940's were profitable years for the industry. It wasn't until the 1950's that the industry hit its first growth challenge: the automatic home washer and dryer. These new machines allowed consumers to clean more of their clothes at home. The 1960's dealt the biggest blow to the industry with the introduction of wash and wear clothing. In the 1970's the popularity of easily cleaned polyester fabrics meant that even fewer clothes were being taken to the cleaners.

The coin-operated and discount dry cleaner arrived to challenge neighbourhood businesses in the 1980's. In the 1990's, casual office wear has become more acceptable. This has caused the market for dry cleaning services to shrink again. Although casual office wear reduces the market for traditionally dry cleaned garments, it also provides a new marketing opportunity for wet cleaners. The dry cleaning industry is also being affected by environmental regulations that have been introduced since the 1970's. These regulations are driving the industry to develop cleaning processes that do not use hazardous chemicals. Several new cleaning processes are currently being researched. The goal of this research is to create new processes that do not harm the environment, but are also efficient, effective and profitable.

These new processes include cleaning with liquid carbon dioxide, ultrasonics, ozone, and wet cleaning. At the time this manual was written, wet cleaning was the only practical, environmentally sound and potentially profitable new method. Therefore, the focus of this manual is on wet cleaning. In this country, 80 per cent of professional cleaners use *perchloroethylene* (or "perc") as their dry cleaning solvent and the remaining 20 per cent use petroleum. Because most cleaners use perchloroethylene, this manual compares wet cleaning with perchloroethylene dry cleaning.

A BRIEF HISTORY OF WET CLEANING

In May 1992, an International Roundtable on Pollution Prevention and Control in the Dry Cleaning Industry was assembled by the United States Environmental Protection Agency (EPA) and its Design for the Environment Programme (DfE). The roundtable was created to consider the health and environmental concerns associated with perchloroethylene dry cleaning. The roundtable participants included professionals from the dry cleaning industry and allied trades, researchers, environmentalists and government officials.

These participants formed a partnership that evaluated many methods for reducing the risks to garment care workers and the environment that are associated with the use of perchloroethylene.

In November and December of 1992, the EPA's Design for Environment (DfE) staff conducted a test study of *multiprocess wet cleaning* at the New York School of Dry Cleaning in New York City. In multiprocess wet cleaning, garments are inspected and pre-treated if necessary. The garments are then spot cleaned, steam cleaned or hand washed.

In this test, the cost and effectiveness of wet cleaning was compared to perchloroethylene dry cleaning. The results of the test are described in the EPA's report "Multiprocess Wet Cleaning— Cost and Performance Comparison of Conventional Dry Cleaning and an Alternative Process," EPA 744-R-93-004, September 1993.

The results stated that under certain situations, multiprocess wet cleaning can be technically feasible and economically competitive with perchloroethylene dry cleaning.

One possible problem this study identified was that multiprocess wet cleaning may require more manual labour than dry cleaning. Additionally, the long term effects on garments were not evaluated.

For this reason, dry cleaners may not want to use the multiprocess wet cleaning method. Since the test was done, however, new wet cleaning machines have become readily available. The wet cleaning machines reduce the amount of cleaning that needs to be done by hand. This makes *machine wet cleaning* more efficient than multiprocess wet cleaning, and more attractive to dry cleaners.

A second problem related to wet cleaning has been Federal labeling requirements. Current United States Federal Trade Commission (FTC) care labeling rules discourage dry cleaners from wet cleaning garments labeled "dry clean only.

" The FTC published two requests for public comment on their labeling rules, one in June, 1994, and another in December, 1995. The FTC wanted to know if it should change these rules in order to (among other things) help the EPA's goal of pollution prevention in the dry cleaning industry. These rules are currently under review.

The DfE roundtable was also given the job of developing pollution prevention awareness and practices in the dry cleaning industry. One of the products of this group is the Cleaner Technology Substitutes Assessment (CTSA).

This study is designed to evaluate garment cleaning methods that could be used instead of perchloroethylene dry cleaning. The study looks at these alternative cleaning methods from environmental, health, and economic points of view. Another role of the study is to communicate information about alternative cleaning methods.

In order to gather information for the study, a demonstration project has been established with partial funding from the EPA. The goals of the demonstration are to examine some of the issues and concerns surrounding garment wet cleaning. This project has resulted in a partnership between a private entrepreneur and the Center for Neighbourhood Technology (CNT) in Chicago. CNT is a Chicago- based non-profit organization dedicated to environmentally sound, local economic development. The demonstration involves a garment care shop, "The Greener Cleaner," which opened to the public in May 1995.

"The Greener Cleaner" uses machine wet cleaning. A garment wet cleaning business depends on customer acceptance in order to be successful. CNT evaluated "The Greener Cleaner" by using customer survey cards and by having volunteers evaluate how well garments were cleaned. Complete results of the CNT study can be obtained by contacting CNT directly.

Wet cleaning technologies have proven to be viable in Europe. In the past few years, these wet cleaning technologies have begun to spread in the United States.

Many commercial cleaners use wet cleaning as a critical supplement to their cleaning process, and there are a growing number of wet clean-only facilities across the U.S. The USEPA's Design for Environment programme has a brochure on wet cleaning which list wet cleaners nationwide.

WHAT DOES THAT MEAN

Multiprocess wet cleaning: A wet cleaning process in which garments are inspected and pre-treated if necessary. The garments are then spot cleaned, steam cleaned or hand washed.

Machine wet cleaning: A process for cleaning textiles in water by professionals using special equipment and detergents. Machines carefully control mechanical action, temperature and humidity. It is normally followed by restorative finishing procedures.

Organic solvent: A solvent that contains carbon in addition to other substances.

Perchloroethylene: An organic solvent frequently used in dry cleaning and as a spot cleaner. Other names for perchloroethylene are: tetrachloroethylene, PCE, and perc.

Pollution prevention: Solving pollution problems by preventing the pollution from occurring in the first place. This is frequently done by eliminating or reducing the amount of materials, hazardous or otherwise, that are used in a process, so that less pollution is created. Elimination is the preferred method of pollution prevention.

Solvent: A liquid that is able to dissolve other materials. Water is the most common solvent. Most solvents are only able to dissolve certain types of materials.

HEALTH, SAFETY AND ENVIRONMENTAL REGULATIONS

HAZARDS

Most of the regulations that are currently in effect are designed to limit the amount of perchloroethylene that comes in contact with workers and the environment. Perchloroethylene, also known as *perc*, can be a hazard to workers and to the environment if it is not used and disposed of properly.

Acute and Chronic Exposure

There are two types of exposure to perc: acute and chronic. Acute exposure to perc occurs when people are exposed to perc for a short amount of time (under 24 hrs). Acute exposure to perc may cause drowsiness, nausea, and irritation of the eyes, skin and respiratory tract.

Chronic exposure may occur when people are exposed to perc repeatedly over a long period of time. Garment care workers have the biggest risk of chronic exposure, because they can be repeatedly exposed to perc at work. The public can also be at risk, however, when perc from a dry cleaning plant gets into the local air and water. Chronic exposure to perc can cause central nervous system damage, and to a lesser degree, lung, liver and kidney damage. The central nervous system effects include irritability, headaches, fatigue and *vertigo*.

There is ongoing debate about whether or not perc is a *carcinogen*, or cancer-causing substance. Different agencies have taken different positions on this question. The U.S. Environmental Protection Agency (EPA) considers perc a suspected carcinogen. The National Institute for Occupational Health and Safety (NIOSH) recommends that perc be handled as a human carcinogen. The International Agency for Research on Cancer (IARC) classified perc as a probable carcinogen in 1995.

Exposure to Workers and Others

Workers are at an especially high risk of occupational exposure in shops where "transfer machines" are used. Transfer machines are dry cleaning machines where the washing and drying take place in two physically separate machines.

After the wash cycle is finished, workers have to manually transfer clothes saturated with perc from washers to dryers.

During this transfer, workers may be exposed on the surface of the skin, such as arms and hands, when their skin comes in contact with the wet clothes. This is called *dermal exposure*. Even more important, workers may be exposed to perc when they breathe in the fumes while transferring the clothes. This is called *inhalation exposure*.

Currently, there are about 10,000 transfer machines in use. Carbon adsorbers and refrigerated condensers are used to control *emissions* and are currently installed in about half of the transfer machines in operation. Newer machines do not require workers to physically transfer clothes from the washer to the dryer.

These newer "closed loop dry-to-dry" machines also have many safety features that help prevent perc from coming in contact with workers or the environment. Dry-to-dry machines allow a dry cleaner to process a load of garments without coming in contact with perc-saturated clothes. Transfer machines can no longer be sold in the U.S., so all the dry cleaning equipment currently available includes these new controls.

People who work in dry cleaning shops have the highest risk of exposure to perc. People living near or above these shops who spend a great deal of time at home, such as the elderly or pregnant women, also can be exposed to perc for extended periods of time.

Environmental Effects

The use of perc, regardless of the type of equipment used, may result in some emissions to the air, water or soil. Perc evaporates quickly when released to soil or surface water (rivers, lakes, oceans, etc). When exposed to sunlight, it breaks down rapidly. When perc gets into water, it does not significantly affect organisms that live in the water. In general, it is not absorbed by aquatic organisms and does not tend to build up in their bodies. A small percentage of the perc in water, however, will be absorbed by microorganisms and enter the food chain.

Currently, there are over 30,000 dry cleaners in the United States, 80 per cent of whom use perc. The total amount of perchloroethylene used for dry cleaning in this country in 1994 was estimated at over 11 million gallons (approximately 147 million pounds).

After it has been used, perc may end up as:

- Emissions into air
- Emissions into water and soil
- Hazardous waste

Perc is one of the ten most common contaminants found in groundwater. A Federal survey reported that perc, from all commercial and industrial sources, is present in about 26 per cent of U.S. groundwater supplies. The California Air Resources Board estimates that 30 per cent of the total perc used—3.3 million gallons—ends up as hazardous waste.

It is estimated that recyclers can recover up to 80 per cent of the waste generated—2.64 million gallons. Although these recovery figures are impressive, the fact remains that hundreds of thousands of gallons of waste are generated each year which could potentially contaminate the environment.

Environment, Health and Safety in Wet Cleaning

Because perc is not used in the wet cleaning process, wet cleaners do not need to worry about perc related environmental health and safety problems. Wet cleaning uses water as the solvent. The detergents, stain removal agents and finishes used in wet cleaning are generally non-toxic. Companies that produce wet cleaning detergents, stain removers and finishes are sensitive to environmental and *toxicological* concerns and are producing products that are phosphate free and *biodegradable.*

Data collected by the Center for Neighbourhood Technology at the Greener Cleaner show that wet cleaning does not produce hazardous waste or air emissions. Workers, customers and people who live and work near a dry cleaner might benefit from the introduction of wet cleaning and the resulting decrease in the use of perchloroethylene. One environmental health and safety concern that does apply to wet cleaning is the use of toxic solvents during stain removal. A new shop with a 100 per cent wet cleaning operation would probably not use toxic stain removal agents, but a shop that has used perc and converted totally or partially to wet cleaning may still use these chemicals at the stain removal board. Non-toxic stain removal agents are currently available from several different sources and should be used on garments that will be wet cleaned. If the plant operator or owner continues to use toxic stain removal agents, the waste generated by the process should be treated as hazardous waste.

WHO MAKES THE RULES?

The federal Occupational Safety and Health Administration (OSHA) and state labour departments are in charge of ensuring the safety and health of workers inside the plant. The federal Environmental Protection Agency and the state departments of environmental protection are in charge of regulating pollutants emitted or transferred from the plant. This is because the legislature and regulatory agencies have always considered worker health and safety issues to be separate from environmental protection issues. Some state agencies are starting to combine environmental protection with worker health and safety. This has resulted in state programmes to reduce the use of toxic chemicals in processes. However, the regulations that are currently in effect are still mainly divided into environmental regulations and health and safety regulations.

The minimum compliance requirements of a dry cleaner based on the Federal regulations. Individual states can make regulations of their own that are even stricter than the Federal regulations. When states make their own regulations, the garment cleaning industry only needs to report to and follow the rules of that state. For example, New York and California are two states that have decided to make their own regulations. Anyone who intends to own or operate a professional garment care facility should research state and local regulations to determine which regulations apply in his or her state.

Since 1970, the U.S. Congress has passed several laws designed to protect and improve environmental and public health. The laws set standards for limits of chemical releases to air, water and soil. These laws originally targeted industries that use large quantities of toxic substances. As these industries have come into compliance, the focus has turned towards industries that use small quantities of toxic substances.

WHAT RULES ARE CURRENTLY IN EFFECT?

The federal regulations outlined here apply to dry cleaning operations. Perc is listed as a hazardous substance under the Occupational Safety and Health Act (OSHA), the Resource Conservation and Recovery Act (RCRA) the Clean Water Act (CWA), the Clean Air Act (CAA), and the Comprehensive Emergency Response, Compensation and Liability Act (CERCLA). A dry cleaner that uses perc must comply with these regulations. Complying with these regulations can affect a dry cleaner's profit margins. The costs associated with compliance include:

- Liability insurance
- Permitting fees
- Hazardous waste disposal fees
- Equipment upgrades
- Contaminated site clean up
- Time spent learning and complying with regulations

Shifting to a mix of wet cleaning and dry cleaning can provide partial relief from some of the regulatory requirements imposed on dry cleaners.

For example, some of the Clean Air Act regulations are based on how much of a toxic chemical is used by a facility. Small garment cleaning businesses with a dry-to-dry machine installed before 1991 are required to upgrade their equipment under the CAA if they purchase more than 140 gallons of perc each year. By switching some cleaning to wet cleaning, a business may be able to reduce its purchases below the 140 gallon limit. Once below 140 gallons, the business would be exempt from the upgrade requirement.

Occupational Safety and Health Act

The Occupational Safety and Health Act of 1970 was passed to protect the health of the worker in the workplace environment. The Act is administered by the Occupational Safety and Health Administration (OSHA). The Act sets limits on the amount of toxic materials that workers can be exposed to in their workplaces.

Permissable Exposure Limits (PELs)

Both long-term and short-term exposure limits have been established. These limits are called the Permissible Exposure Limits (PELs). The long-term limits refer to a worker's allowable average exposure during a normal 8-hour workshift of a 40-hour work week. The average exposure limit takes into account how long a worker is exposed to different levels of the chemical. This average is call the Time Weighted Average (TWA). The overall average exposure for any 8-hour workday of a 40-hour work week cannot be greater than the PEL. The current long-term PEL for perc is 100 parts per million (ppm). In 1989, because of concerns about whether the 100 ppm limit is low enough to protect workers, OSHA reduced the limit to 25 ppm. Dry cleaners had until December 31, 1991 to meet the new standard.

In fact, many dry cleaners have installed control equipment to control emissions so that they meet the 25 ppm PEL. However, soon after the 25 ppm standard was issued, a suit was filed by industry groups to challenge the new limits and a court decision threw out the 25 ppm limit. As a result, currently OSHA can only enforce a PEL of 100 ppm but still recommends that dry cleaners limit indoor air emissions of perc to 25 ppm. Several state occupational agencies continue to enforce the 25 ppm limit. In addition, the American Conference of Governmental Industrial Hygienists (ACGIH) and the perc manufacturers also recommend 25 ppm as an 8-hour TWA. Currently, negotiations are under way to reduce the PEL again, with some public interest groups recommending levels as low as 5 ppm.

In addition to the long-term PEL, two short-term exposure limits have been established by OSHA. These are designed to protect workers from high exposures that might occur occasionally throughout the work day. In dry cleaning, this might occur when opening the machine to remove the clothes, when cleaning the still or filter housings, when cleaning out the button trap, or at any other time when perc vapors are in close contact with the worker.

The OSHA short-term PELs for perc are 200 ppm averaged over 15 minutes, and 300 ppm averaged over 5 minutes. The 5 minute maximum (peak) exposure can only occur once in any 3 hour period. ACGIH recommends a short-term exposure limit of 100 ppm, averaged over 15 minutes.

Resource Conservation and Recovery Act

Subtitle C of the Resource Conservation and Recovery Act (RCRA) of 1976 regulates what is done with hazardous waste that is generated at facilities. The regulations only apply to facilities that generate more than 100 kilograms (220 pounds) per month of hazardous wastes. Facilities that generate less than this amount are considered small quantity generators and are exempt from this regulation. Waste that contains perc is one of the hazardous wastes regulated under RCRA. In the case of a dry cleaner, this waste could consist of still bottoms, cartridge filters and filter muck. This waste must be disposed of at a licensed hazardous waste facility.

Some water used in the dry cleaning process becomes contaminated with perc. This occurs when water is used by emission control devices and perc filtration devices, and when water is used to maintain these devices. Waste water generated by dry cleaners, even though only slightly contaminated, is considered hazardous waste under RCRA. The hazardous waste code for this type of waste is F002. Separator water is also considered hazardous waste because it may contain as much as 150 ppm of perchloroethylene. Solid wastes and waste water from wet cleaning processes are not considered hazardous under RCRA. Some stain removal agents may also generate hazardous waste.

Clean Air Act (1977) and Clean Air Act Amendments (1990)

The Clean Air Act of 1977 (CAA) and the Clean Air Act Amendments of 1990 (CAAA) also affect dry cleaners. The CAAA added perc to the list of regulated hazardous air pollutants. In September 1993 the CAAA put out the National Emission Standards for Hazardous Air Pollutants (NESHAP), which regulates emissions of perc by dry cleaners. These standards limit the amount of emissions of regulated hazardous air pollutants including perc.

The CAAA requires that dry cleaners keep track of their use of perc and its disposal. It requires dry cleaners to change their operating procedures and modify their existing equipment to meet new air emissions standards. It also requires that all newly purchased dry cleaning machinery be equipped with control technology. These standards also prohibit the sale of transfer machines. Federal law requires that the industry be in compliance with the perc NESHAP by September 23, 1996.

New Source Performance Standards (NSPS) were put into effect in 1984 for petroleum based dry cleaners. The NSPS's are enforced only in Clean Air Act non-attainment areas—those areas that do not meet national standards for ambient air quality. The NSPS's set limits on solvent loss in drying, set standards on the use of filters, and require leaks to be repaired in a timely fashion. Wet cleaning processes do not emit hazardous air pollutants that are regulated under the Clean Air Act.

Comprehensive Emergency Response, Compensation and Liability Act (1980) and Superfund Amendments and Reauthorization Act (1986)

The Comprehensive Emergency Response, Compensation, and Liability Act of 1980 (CERCLA), commonly referred to as Superfund, allows the EPA to respond to potential or actual releases of hazardous substances that threaten public health, welfare or the environment. Superfund also authorizes the EPA to require parties responsible for environmental contamination to clean up or pay back the Superfund for costs incurred by EPA to clean it up.

The CERCLA hazardous substance release reporting regulations (40 CFR Part 302) require the person in charge of a facility to report certain releases of hazardous substances. If the amount released is greater than the limit specified in the regulation (40 CFR § 302.4), then the person in charge must contact the

National Response Center (NRC). This may trigger a response by the EPA or a number of state or Federal emergency response agencies. When a site is found to be contaminated, present or past landowners or dry cleaners can be held liable for any contamination found.

The contamination may have resulted from a variety of situations, such as:

- Contaminated waste water leaking through sewer pipes
- Perc leaks during operation of a dry cleaning machine
- Dumping of wastes

The materials used in wet cleaning are not a source of liability under the CERCLA.

Sewer or Septic Discharge

Water discharge from a dry cleaning or wet cleaning facility is subject to restrictions depending on the size of the facility and whether the discharge is to a publicly owned treatment facility or to a septic system.

Specifically, the following restrictions apply:

- Contact water (separator and vacuum water) must be disposed of properly as hazardous waste or may be evaporated under an EPA policy decision
- Boiler blowdown water is regulated on a state by state basis
- Vacuum water from the finishing equipment must be disposed of as hazardous waste
- Waste from the stain removal board must be treated as hazardous waste

The waste water generated by a wet cleaning operation is discharged either to a municipal sewer—Publicly-Owned Treatment Works or POTW—or to a private septic system. Cleaners should always check with their local POTW, state agency or municipality before discharging into either a sewer or septic system since they may require that waste water must be non-toxic in order to avoid polluting the environment. This will ensure that it is legal to discharge their waste water and that no permits are required. For average size wet cleaning facilities, a discharge permit, a pretreatment system, or monitoring of discharge may not be required by the Publicly- Owned Treatment Works.

Discharge to septic systems is regulated by the local municipalities and the state. Waste from a professional cleaner is considered industrial waste, and, in many states, discharge into a septic system is prohibited. The local regulating authorities, appropriate state agencies and the POTW should be notified during the planning stages of any wet cleaning or dry cleaning facility to determine any applicable regulations or permitting requirements.

WHAT ARE THE POTENTIAL LIABILITIES?

When new regulations are passed that restrict dry cleaning facilities, the public becomes more aware of the public health and environmental risks of perc. The professional fabricare industry, primarily composed of very small "mom and pop" operations, has been hard hit by these new regulations.

Landlords are wary of renting to dry cleaners because of the potential liability associated with perc. The cost of compliance and the potential for retroactive liability has affected the profit margin of many operations, forcing some owners to go out of business. Some families that in the past might have passed the family dry cleaning business down to the next generation must now be mindful that current retroactive liability laws may transfer past liability to the new owner.

These regulations have had far-reaching effects on the dry cleaning industry. Not only do dry cleaners need to comply with new Federal regulations by implementing expensive pollution control technology, they need to consider that their home state could pass even stricter regulations in the future. A cleaner who makes a partial or total switch to machine wet cleaning (or water based cleaning) is subject to limited regulations, and in some cases, no regulations at all.

What does that Mean?

Acute exposure: contact with a substance that occurs over a short period of time

- *Biodegradable:* Able to be broken down into non—harmful substances by natural processes in the environment, such as sunlight, rain and wind
- *CAA:* Clean Air Act of 1977
- *CAAA:* Clean Air Act Amendments of 1990
- *Carcinogen:* A substance that causes cancer
- *CERCLA:* Comprehensive Emergency Response, Compensation, and Liability Act
- *Chronic exposure:* Contact with a substance that occurs over a long period of time
- *CWA:* Clean Water Act
- *Dermal exposure:* Exposure to a substance when the substance comes in contact with the skin
- *Emission:* The release of a substance into the air, water or ground
- *Exposure:* Coming in contact with a substance. The contact can be on the skin, through breathing into the lungs, or by eating or drinking.
- *Inhalation exposure:* Exposure to a substance that occurs when something in the air is breathed into the lungs
- *NESHAP:* National Emission Standards for Hazardous Air Pollutants
- *NSPS:* New Source Performance Standards
- *OSHA:* Occupational Safety and Health Administration; the part of the Federal government that administers the Occupational Safety and Health Act of 1970

- *PEL:* Permissible Exposure Limit; the maximum amount of a substance that a worker is allowed to come in contact with based on a daily or weekly average
- *Perc:* A short name for perchloroethylene
- *RCRA:* Resource Conservation and Recovery Act of 1976
- *Toxicology:* The science that studies the effects, antidotes and detection of poisons or adverse effects of chemicals on living organisms.
- *SARA:* Superfund Amendments and Reauthorization Act
- *Superfund:* Another name for CERCLA
- *TWA:* Time Weighted Average; the method used to calculate a worker's average exposure to a substance
- *Vertigo:* Dizziness

FIBRES, YARNS, AND FABRICS

A customer walks into a cleaning shop and puts a garment on the counter. To the customer, the garment is only a shirt, or a pair of pants, or a dress. When a cleaner looks at that same garment, he or she sees much more than that. The cleaner sees that the garment is made from a particular type of fabric, and the fabric is made from a particular type of fibre.

To successfully wet clean the garment, the cleaner must understand how the fabric and fibre will react to the cleaning process. Fabrics and fibres react differently in water than they do in perchloroethylene or other dry cleaning solvents. Even an experienced fabricare professional must pay careful attention to fabric and fibre when wet cleaning.

This module describes the characteristics of fibres and fabrics. This module also explains the effect of the cleaning process on fibres and fabrics.

WHAT IS A FIBRE?

A fibre is a fine hair-like structure that is either natural or man-made. Natural fibres come primarily from plants and animals. Plant fibres come from the seeds of plants like cotton, or from stems like linen, ramie, hemp, or jute. Animal fibres come from the hair of certain animals. For example, the hair of sheep is used to make wool and the hair of goats is used to make cashmere and mohair. Silk fibres are spun by silkworms.

Manufactured fibres, also called man-made fibres, were developed to provide properties that natural fibres do not have. Manufactured fibres are divided into two main categories: synthetic fibres and regenerated fibres. Synthetic fibres are created from polymers that come from chemicals like petroleum. Nylon, polyester and acrylic are all synthetic fibres. Regenerated fibres are created from polymers that come from natural materials. Rayon and acetate are both regenerated from *cellulose*, a material that occurs naturally in plants.

Manufactured fibres are made using a process that is very similar to the process a spider uses to make strands for its web. Long chains of molecules called *polymers* are made into a liquid. This liquid is forced through tiny holes. As the liquid comes out of the holes, it hardens into fibre strands called *filaments*, or strands of "endless" length.

Both natural and manufactured fibres can be made directly into fabric, such as felts and nonwoven textiles. However, fibres are usually twisted or grouped together into long strands called yarns. The yarns are then used to make a wide variety of knit and woven fabrics, as well as lace.

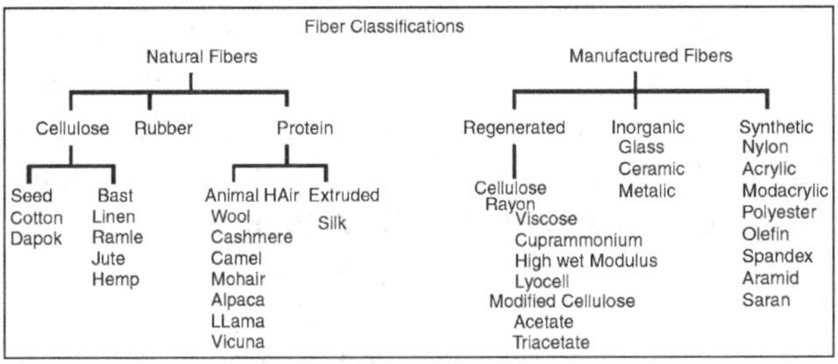

Fig. Fibre Classifications.

The length of the fibre helps to determine the properties of the yarn and of the fabric that is made from that yarn. Almost all natural fibres range in length from less than one inch up to 36 inches. These are called *staple fibres*. The exception to this rule is silk. The silk worm spins filaments of silk that can be up to 1,600 yards long. Like silk, all manufactured fibres are also spun as filaments. Unlike silk, there is no limit to the length of manufactured fibres. However, these fibres can be cut into short or "staple" lengths if necessary.

YARNS

Yarns are used to make a wide variety of knit fabrics, woven fabrics and lace. The type of fibre used to make a yarn and the way the yarn is made determine the properties of the yarn and of the final fabric. These properties include strength, *abrasion resistance, elasticity, luster, hand* and overall surface characteristics.

In many cases, fibres are twisted to make yarn. Increasing the twists or turns in a yarn will increase strength, crease resistance and abrasion resistance. Yarns that are more loosely twisted make softer fabrics.

Some yarns are made from a single type of fibre, while other yarns are made from more than one type of fibre. Multiple ply yarns are stronger than single yarns. Fancy yarns, like boucle or chenille, can be made to create decorative effects, and may take special care in cleaning. Figure shows several different types of fibres and yarns. Filament fibres are typically sleek and smooth. The surface of fabric made from filament fibres is therefore also sleek and smooth.

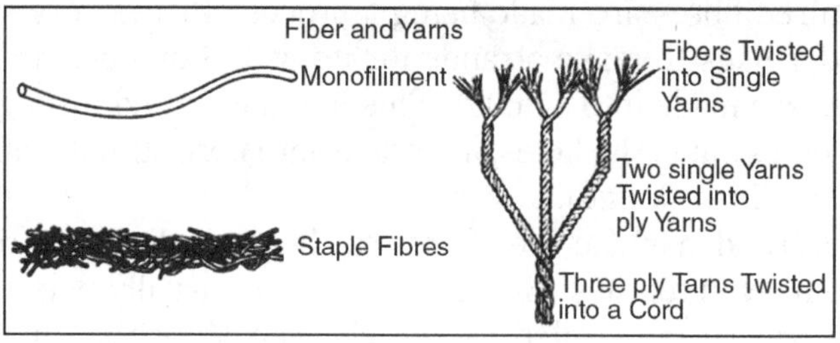

Fig. Fibre and Yarns.

Natural staple fibres tend to be rougher. The surface of fabric made from staple fibres will also be rougher, due to the exposed fibre ends. For example, nylon and silk fibres have round, smooth surfaces, and therefore nylon and silk fabric may feel smooth and slippery. Wool fibres have a round scaly surface, making wool fabrics feel bulky and slightly rough.

FABRIC TYPES

The basic fabric types can be classified as woven, knitted, lace and non-woven. Woven fabrics consist of two sets of parallel yarns that are interlaced to form a fabric. The lengthwise yarns are called "warp." Warp yarns are held under tension throughout the weaving process. The crosswise yarns are called "weft" or "filling." Some examples of woven fabrics are plain, twill and satin weaves, as well as certain pile fabrics like velvet, velveteen and corduroy. Figure shows some examples of woven fabrics.

The weaving process tends to create fabrics that are strong and do not lose their shape. The heavier the fabric and the tighter the weave, the more durable the fabric will be. This assumes, of course, that the yarn and fibre are also strong and stable. Fabrics that are tightly woven with delicate yarns, for example, can still be very fragile. Knitted fabrics consist of one or more continuous yarns that are looped through themselves to form interconnected chains. Weft knits are the most common type of knits. Weft knits use one continuous yarn to form the fabric through a series of loops, and are easy to snag and run. Warp knits use a series of continuous yarns to form the fabric, giving the fabric more stability and less tendency to run. Figure shows examples of warp and weft knits.

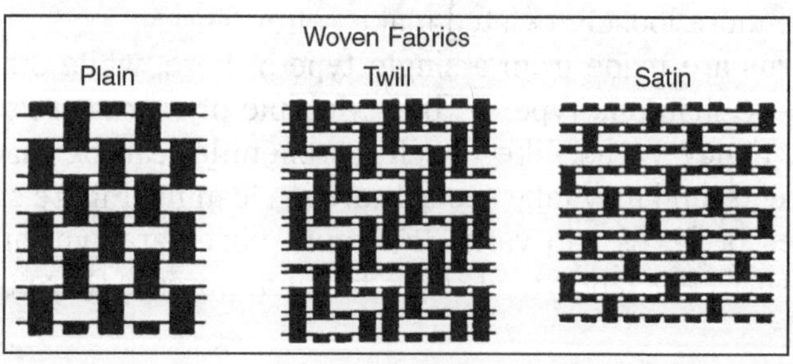

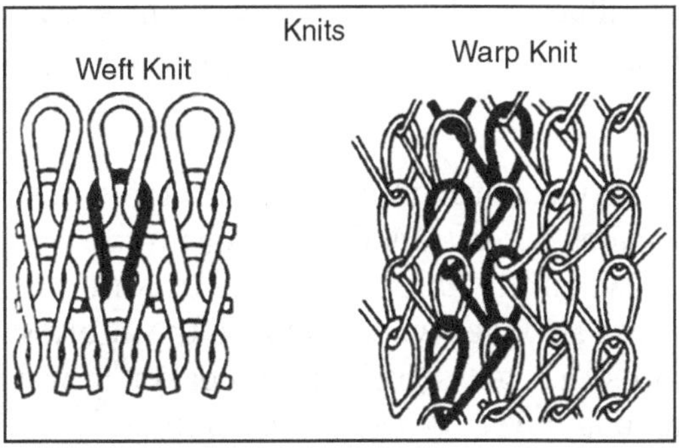

Knit fabrics tend to be more comfortable than woven fabrics because they stretch. This is an advantage for the person who wears the fabric. It can cause problems for the cleaner, however, because this tendency to stretch can cause the fabric to lose its shape when it is handled. Knitted fabrics also tend to shrink. Fortunately, knitted fabrics can be 'blocked' or stretched to recover their original shape and size.

Because some fabrics are constructed of loosely twisted yarns, pilling is often a problem. When a knitted fabric is rubbed repeatedly, small balls of fibre called "pills" build up on the surface of the fabric. This is called "pilling." Cleaners need to be careful not to cause pilling when they rub the surface of a knit fabric to remove soil or stains.

Lace is made of yarns that are knotted, twisted, or looped to provide a fragile, sheer pattern. When cleaning garments with lace, great care must be taken not to break the delicate yarns. Non-wovens are fabrics made by bonding together a web of fibres through physical or chemical means. Felt is an example of a non-woven fabric. Felt is made primarily from wool fibres. The felting process uses moisture, heat, mechanical action and chemical action to take advantage of the natural tendency of wool fibres to stick together. The result is a dense, bulky fabric. Imitation suede is another example of a non-woven fabric.

Non-woven fabrics tend to be weaker and more easily damaged by rubbing than woven or knit fabrics of the same weight. This is because non-woven fabrics are made directly from fibres, and therefore do not benefit from the strength of yarns. Non-woven fabrics are often used for interfacing. Non-wovens can be bonded to a fabric to make it more stable. This can cause problems for the cleaner. One problem is that the cleaning process can dissolve the adhesive and cause the two fabrics to separate, or "delaminate." Another problem is that one fabric may shrink more than the other, which can cause puckering.

THE EFFECT OF CLEANING ON FIBRES

Each type of fibre reacts differently to the wet cleaning process. How a fibre reacts depends on a number of factors, including strength, elasticity, dimensional stability, absorbency and the effect of chemicals and heat. These factors have a clear influence on the way a garment should be handled in the wet cleaning process.

Fibre Strength

Fibre strength is a measure of the amount of force needed to cause a fibre to break. It is also called *tenacity*, and it is measured in grams/ denier. Fibre strength can change when the fibre is exposed to water. Cotton and linen are the only fibres that get stronger when exposed to water. Rayons and acetates are examples of fibres that get significantly weaker in water. Polyesters and olefins, which are among the strongest of the common fibres, do not get weaker or stronger. The strength of these fibres is not affected by water at all. Figure shows the wet and dry strength of different fibres.

The effect of water on fibre strength is very important for the wet cleaner in the stain removal and cleaning stages of garment care. Because the strength of fibres can change when the fibres are wet, care must be taken when applying any force to wet fibres. For example, rayons and acetates, since they get significantly weaker in water, could be damaged by scrubbing or harsh agitation. Figure shows the resistance of different fibres to abrasion.

	Dry (Grams/Denier)	Wet (Grams/Denier)
Acetate	1.2 - 1.5	0.8 - 1.2
Acrylic	2.0 - 3.5	1.8 -3.3
Cotton	3.0 -5.0	3.3 - 6.4
Linen	Stronger than Cotton	
Modacrylic	2.0 3.5	2.0 - 3.5
Nylon	3.0 -6.0*	2.6 - 5.4
Olefin	4.8 - 7.0	4.8 - 7.0
Polyester	4.0 - 5.0*	4.0 - 5.0
Rayon	0.73 - 2.6*	0.70 - 1.8
Silk	2.4 - 5.1	1.8 - 4.2
Wool	1.0 - 1.7	0.8 - 1.6

* High Tenaclty Nylon High as 9.5 g/den
High Tenacity Polyester as High as 8.0 g/den
High wet modulus Rayon- 4.5 g/den

Fig. Fibre Strength.

Elasticity and Dimensional Stability

Elasticity is the ability of a fibre to go back to its original length after being stretched. Fibres that have good elasticity are those that return to their original length when stretched. Cotton, linen and rayon have poor elasticity, while wool, nylon, polyester and olefins are very elastic fibers.

The *dimensional stability* of a fiber is its ability to maintain its original shape. Fibers with good dimensional stability do not shrink or stretch during cleaning. Fibers with poor dimensional stability may shrink or stretch during cleaning.

Abrasion Resistance

Fiber	Abrasion Resistance
Acetate	Fair to low
Acrylic	Good to Sufficient
Cotton	Good*
Linen	Good, but Damaged by Repeated Flexing
Nylon	Good to Excellent
Polyester	Good to Excellent
Rayon	Poortofair
Silk	Poor
Wool	Good*

* Wool and cotton are rated or lower than each other in differents situations

Wool and rayon fibers have poor dimensional stability. They tend to shrink if not treated with care when exposed to water. When a rayon garment shrinks, it can often be stretched back to its original shape in the finishing process. However, when a wool garment shrinks, it can be difficult to return it to its original shape. Proper care when cleaning garments can prevent shrinkage. For example, wool fabric should never be cleaned using moisture with heat, or moisture with vigorous agitation, because these combinations can cause felting and thus permanent shrinkage.

Even some fibers with good dimensional stability can lose their shape during the cleaning process. Synthetic fibers tend to have good dimensional stability, but exposure to high temperatures can permanently change their shape. Figure shows the elasticity and dimensional stability of common fibers.

Absorbency and the Effect of Water on Fibers

A fiber's ability to take up moisture is referred to as absorbency. In technical terms, absorbency is called *moisture regain*. Moisture regain is a measure of the weight of the moisture a fiber absorbs compared to the weight of the fiber itself. It is measured at a standard temperature and humidity, usually 70??F and 65 per cent humidity.

If 100 grams of a fiber absorbs 5 grams of moisture, then the fiber has a moisture regain of 5 per cent. Cotton, for example, has a high moisture regain of 7 per cent. One hundred grams of cotton will absorb 7 grams of moisture. Polyester, on the other hand, has a low moisture regain of only 0.2 per cent. One hundred grams of polyester will only absorb 0.2 grams of water. Figure shows the standard moisture regain for common fibers.

Fibers with a high moisture regain are also called *hydrophilic* or "water-loving" fibers. These fibers possess the ability to absorb water with ease. All natural fibers, as well as rayon and acetates, are examples of hydrophilic fibers. These fibers tend to swell when exposed to water.

The fact that these fibers tend to absorb water is both an advantage and a disadvantage to the cleaner. The primary advantage is that stain removal is easier, because water and detergent are easily absorbed into and flushed out of the fiber. The primary disadvantage is that swollen, saturated fibers can also cause increased shrinkage, wrinkling and loss of shape of the garment.

Elasticity and Dimensional Stability

	Elasticity	Dimensional Stability
Acetate	Poor	Fair
Acrylic	Good	Poor
Cotton	Poor	Good
Linen	Poor	Good
Modacrylic	Fair	Good if not Exposed to High Temperatures
Nylon	Excellent	Good if not Exposed to High Temperatures
Olefin	Excellent	Good if not Exposed to High Temperatures
Polyester	Excellent	Good if not Exposed to High Temperatures
Rayon	Poor	Poor
Silk	Good	Good
Wool	Excellent	Poor

Fibers with a low moisture regain do not absorb water or absorb only small amounts of water. These fibers are referred to as *hydrophobic* or "water-fearing." Synthetic and inorganic fibers such as polyester and nylon are hydrophobic. These fibers tend to wrinkle less when processed in water. In fact, synthetic fibers keep their original shape when processed in water, just as natural fibers tend to keep their shape when processed in dry cleaning solvents.

Absorbency

	Moisture Regain (%)*
Acetate	6.0 - 6.5
Acrylic	1.0 - 2.5
Cotton	7.0 - 11.0
Linen	8.0 - 12.0
Modacrylic	0.4 - 4.0
Nylon	4.0 - 4.5
Olefin	0.01 - 0.1
Polyester	0.2 - 0.8
Rayon	11.0 - 15.0
Silk	11.0
Wool	13.0 - 18.0

* As a Persentage of the dry Weight at 70.F and 65% Relative Humidity

The majority of garments today consist of blends of natural and manufactured fibers. Because synthetic fibers do not tend to be affected by water, these garments can be easier to clean than garments made entirely of natural fibers.

Some natural fibers are treated with special processes to make them easier and safer to clean. Two of these processes are mercerizing and Sanforizing™. Mercerizing is a process in which cotton fibers, yarns or fabrics are held under tension in a cold alkali bath. This strengthens the cotton fibers, makes them shinier, and allows them to absorb dye better. The Sanforizing™ process is a pre-shrinking process. This means that the fabric manufacturer shrinks the fabric on purpose. Once the fabric is shrunk, it is sent to a clothing manufacturer to be made into garments. The result is that the garments will not shrink more than two per cent when washed and dried in home machines.

Effect of Chemicals

A number of different chemicals are useful for getting out soils and stains, and for identifying fabric types. However, these same chemicals can also damage certain fibers. It is important for the garment care professional to understand the effect of certain chemicals on fibers. These chemicals include bleaches, alkalies, acids, and organic solvents.

Bleaches: Bleaches are used to remove colour or stains from fabrics. There are two types of bleaches commonly used in fabric care: *oxidizing bleaches* and *reducing bleaches*.

Oxidizing bleaches include hydrogen peroxide (H2O2), sodium perborate, and sodium hypochlorite (NaClO), also known as chlorine bleach. Reducing bleaches include sodium bisulfite, sodium hydrosulfite and titanium stripper. Of all the bleaches used in garment cleaning, chlorine is the most harmful to fibers. Most other bleaches could be used safely.

Acrylics and polyesters are resistant to damage from all types of bleaches. Cotton, linen, and rayon fibers are not harmed by reducing bleaches, but can only withstand weak chlorine bleaches for short periods of time. Wool and silk—and animal fibers in general— will deteriorate and dissolve in chlorine bleaches, but are more resistant to other bleaches. Some synthetic fibers, such as nylon and modacrylic, may also be harmed by chlorine bleaches. Figure shows the resistance of different fibers to bleaches.

Acids and Alkalies: *Acids* and *alkalies* are used in laboratories for fiber identification and in cleaning to remove difficult stains like rust or blood. As might be expected, acids and alkalies each have different effects on fibers. Synthetic fibers are generally resistant to alkalies and acids, with some exceptions. Acrylic and polyester are sensitive to strong alkalies at high temperatures. Acetate, nylon 66 and polyester will decompose in strong acids.

The natural fibers are affected differently by acids than alkalies. Plant fibers that contain cellulose are sensitive to acids, but are typically not damaged by alkalies. In fact, exposing cotton to alkalies strengthens the cotton fibers. This is the basis of a process called mercerization. The protein, or animal fibers, are generally unaffected by acids, but are damaged by even weak alkalies. Figures show the effect of acids and alkalies on different fibers.

Organic Solvents: Organic solvents, like perchloroethylene, petroleum and acetone, are used to remove a variety of oil-based stains from fabrics. Natural fibers, both animal fibers and cellulose fibers, either resist organic solvents or are unaffected by them. This fact has allowed these fibers to be cleaned in dry cleaning solvents like perc and petroleum with little or no damage to the fibers. Synthetic fibers, however, may be damaged by organic solvents. For example, acetate and modacrylic dissolve in acetone, which is used to remove nail polish and paint stains. The olefin fibers (polypropylene, polyethylene) will dissolve in solvents like perc in temperatures above 160??F. Since temperatures in a perc dry cleaning machine do not get this high, it is possible to dry clean these fibers. Figure shows the effects of organic solvents on different types of fibers.

Effect of Bleaches	
Acetate	Resistant to Oxidizing Bleaches
Acrylic	Highly Resistant
Cotton	Resistent, but chlorine beaches will Destroy if Uncontrolled
Linen	Similar to Cotton
Modacrylic	Some Fibers may be harmed by Chlorine Bleach
Nylon	May be Harmed by Chlorine Bleach
Polyester	Highly Resistant
Rayon	Similar to Cotton
Silk	Deteriorates in Chlorine Bleach
Wool	Will Yellow and Dissolve in Chlorine Bleach

Effect of Acids	
Acetate	Soluble in Acetic acid (33% or Stronger) Decomposed by strong Acids
Acrylic	Resistant to Most Acids
Cotton	Similar to Rayon
Linen	Similar to Rayon
Modacrylic	Resistsnt to most Acids
Nylon 66	Decomposed by Strong Mineral Acid Resistant to Weak Acids
Olefin	Very Resistant
Polyester	Resistant to most Mineral Acids; Disitegreted by 96% Sulfuric
Rayon	Disintegrates in Cold more Sensitive
Silk	Similar to Wool, but more Sensitive
Wool	Generally Unaffected by Acid

There is one major difference between the effects of organic solvents and the effects of water on fibers. The application of some organic solvents to some synthetic fibers, may cause the fibers to melt or dissolve. If this happens, there is no way to reverse this effect.

Effect of Alkalies	
Acetate	Little Effect from Weak Alkalies
Acrylic	Resists weak Alkalies
Cotton	Swells when Treated with Caustic soda, but is not Damaged
Linen	Very Resistant
Modacrylic	Resistant to alkalies
Nylon 66	litter or no Effect
Olefin	Very resistant
Polyester	Resistant to cold Alkalies
Rayon	No effect by week Alkalies; Swells and loses Strength in concentrated Alkalies
Silk	Similar to Wool, but Damage is Slower
Wool	Attacked by weak Alkalies; destroyed by Strong Alkanes

Effecct of Organic Solvents	
Acetate	Soluble in Acetone
Acrylic	May swell in some Solvents
Cotton	Resistant
Linen	Unaffected
Modacrylic	Soluble in Warm Acetone, Otherwise Unaffected
Nylon 66	Generally Unaffected
Olefin	Soluble in Chlorinated Hydrocarbons Above 160°F
Polyester	Generally Unaffected
Rayon	Unaffected
Silk	Unaffected
Wool	Unaffected

When water is applied to natural fibers, the fibers may shrink or lose their shape. Some of these effects can be reversed, and both of these effects can be avoided.

Effect of Heat

The effect of heat on fibers is an important factor in both the cleaning and finishing of garments. A cleaner should ask himself or herself a number of questions before applying heat to a fiber:

- Does this fiber melt or decompose, and if so, at what temperature?
- Does this fiber shrink when it is heated?
- Does this fiber stretch when heated under tension?

Can this fiber be heat-set, or become fixed in a certain state after heating? If one heat this fiber for a long time, will it damage the fiber and make the fabric useless?

Natural fibers do not melt or soften when they are heated. This allows them to be exposed to high temperatures. In fact, fabrics made from natural fibers may actually need these high temperatures to remove wrinkles in the finishing process. Dry heat has very little effect on natural fibers. However, these fabrics may shrink or stretch if exposed to hot water or to moist heat, and must be closely monitored to achieve the desired result. Wools and rayons require particular care in moist heat.

Synthetic fibers have softening or sticking points ranging from 260?F (olefins) to 490?F (acrylic). The lower this softening point, the more care must be taken when using heat in the care of a garment. Too much heat can cause a fabric to stick to pressing equipment or to become permanently heat-set in an undesirable state.

THE EFFECT OF CLEANING ON FABRICS

Wet cleaning can negatively affect fabrics in three main ways:

- Fabrics can shrink or stretch
- Colors can bleed
- Textile finishes may be removed

Understanding these effects can make a cleaner more successful.

Shrinkage

The most common form of shrinkage is called "relaxation shrinkage." During the manufacturing process, fibers and fabrics are constantly stretched in the warp, or length, direction. After manufacturing, the fabric wants to "relax" and return to its natural length. During cleaning, water or solvent can cause the fabric to "relax." This can result in shrinkage.

Relaxation shrinkage usually occurs the first time the garment is cleaned. Woven fabrics generally shrink more lengthwise. This is the direction of the warp yarns when the fabric was woven. Knitted fabrics tend to shrink in both directions.

There are three other main types of shrinkage: natural shrinkage, progressive shrinkage, and differential shrinkage. "Natural shrinkage" depends on the properties of fibers. For example, wool and rayon tend to shrink, while polyesters do not.

Shrinkage in wool is often called "felting." Felting can occur when heat, moisture, and mechanical action are applied to wool fabric. This causes the wool fibers to interlock and results in shrinkage. A "felted" fabric feels thicker and more compact than the original fabric. "Progressive shrinkage" occurs when a fabric shrinks more each time it is cleaned. This is an important point to note, because even if a garment has been cleaned before, it may continue to shrink.

Differential shrinkage may occur when a garment is constructed of two or more fabric types. Each fabric may shrink differently, causing puckering of the seams or baggy linings.

Colorfastness

Colorfastness is the tendency of a fabric to hold or lose its colour. A fabric that is "colorfast" will not lose its colour and "bleed" on other garments during cleaning.

Colour can be added to a polymer, a fiber, a yarn, a fabric, or a garment. Colour is added using dyes or pigments. Different dyes are used for different fibers, providing varying levels of colorfastness. The more completely the dye penetrates the fiber, the better the colorfastness of the fabric will be. Manufactured fibers tend to be the most colorfast. Because the fiber is dyed as a polymer, the colour is deeply imbedded in the fiber structure. When dyeing is done at a later stage in the manufacturing process, colorfastness may suffer. Some fabrics are printed using pigments, not dyes. These pigments sit on the surface of the fiber and may be sensitive to rubbing or abrasion.

Fabric can lose its colour for a number of reasons, including:

- Exposure to water or chemicals such as acids, alkalides, bleaches or perc.
- Exposure to light, especially sunlight
- Exposure to atmospheric gases
- Exposure to heat
- Perspiration
- Abrasion or rubbing

Garments must be tested for colorfastness before they are cleaned. This is especially important for dark fabrics and for garments that are made of both dark and light fabrics. Experienced garment care professionals, no matter what the cleaning method, will learn which types and colors of fabrics are the problems or "bleeders."

Textile Finishes

Finishes are applied to fibers, yarns and fabrics for a variety of reasons. They can be used to modify the appearance of the fabric to provide a shine or soft feel to the fabric surface. Finishes can also be applied to fabric to reduce its shrinkage, help prevent it from wrinkling, or improve its resistance to water or soil. Finishes can be applied to fabric by mechanical or chemical means.

The finish can be:

- Permanent, which means it lasts as long as the fabric lasts
- Durable, which means it lasts for a relatively long time
- Temporary, which means it lasts for a relatively short time

One of the finishes that is used on fabric is called "sizing." Sizing is a chemical treatment that adds body and luster to fabrics. In this process the microscopic spaces between the individual fibers are filled with starch or resin. This reduces the amount of water the fabric will absorbed, which means the fabric will not swell as much when it is cleaned with water.

The starch and gelatin sizings used on cottons and rayons are usually temporary and may be removed by water.When the sizing is removed, fabrics can shrink, lose their shape, or lose their surface finish. Some more durable resin sizings last somewhat longer on the fabric surface, and are not affected as much by water.

GARMENT CONSTRUCTION

Garments are made of many different parts. All these parts must be considered when cleaning the garment. The first part to consider is the fiber and fabric that the garment is made of. A garment can be made from a single-fiber fabric, a multi-fiber fabric (or blend), or more than one type of fabric:

- A garment that is made from only one type of fiber should be cleaned just as to the properties of that fiber. For example, a wool sweater should not be cleaned in hot water with vigorous agitation. A rayon blouse should be treated very delicately and may need some reshaping or stretching during the finishing process.
- A garment made from a fabric blend—such as wool/polyester or polyester/rayon—should be cleaned with the properties of the more delicate or sensitive fiber in mind. For example, a wool/polyester blend should be treated as if it were wool, and a polyester/rayon blend should be treated as if it were rayon.
- Garments that are made of several types of fabrics can present two major problems. If the fabrics are very different colors— such a navy blue and white—the darker colour must be tested for colorfastness before the garment is cleaned. If the fabrics are made of different fibers, shrinkage may occur at different rates in the different fabrics. The garment should be cleaned with the properties of the more delicate or sensitive fiber in mind, to best prevent either fabric from shrinking. In addition to the fiber and fabric, other parts of the garment also need to be considered:
- Care must be taken on certain garments that seams, collars, and lapels do not pucker. These parts of the garment are often made of different materials that shrink at different rates. For example, when the fabric shrinks at a different rate than the thread or fused interfacing, puckering can occur. Often, this puckering can be fixed during finishing, but it is better to avoid the problem than to fix it.
- Delicate garments may be prone to seam ravelling and edge fray, especially if the seams are not secured with proper tailoring. Seams should be inspected before the garment is cleaned in order to determine if this will be a problem.

- Plastic buttons and non-fabric trims are typically a problem for dry cleaners because plastic and non-fabric trims can melt or stick when they come in contact with perc. This is not a problem with the wet cleaning process because water will not damage these types of materials.

HOW TO IDENTIFY FIBERS

Garment care professionals should know how to identify the fiber content of a garment. Knowing the types and per centages of fibers in a garment will help the cleaner determine the best care and treatment. Federal law requires that certain information be provided with each garment, including fiber content and care labelling.

These labels are usually attached permanently to the garment. This is not required by law in all circumstances, however, and consumers will often remove labels for reasons of comfort. If a garment has no identifying label, the cleaner should take the time to identify the garment's fiber content.

The burn test is a simple method of fiber identification. This test can be done on almost any garment where there is an unexposed area from which a small sample can be clipped.

Never try to do the burn test on a piece of fabric that is still attached to the garment. Different fibers have distinct odors and appearances when they are burned. Exact identification can be difficult, but the cleaner will almost always be able to narrow the identification down to one of three categories: cellulose, protein, or synthetic. Based on this information, the cleaner can determine the best method for cleaning the garment. Figure shows the burning characteristics, odor of residue, and appearance for several cellulose, protein and synthetic fibers.

SOILS, ODORS, STAINS AND SEPARATION

SOILS, ODORS AND STAINS

Soils, odors and stains are some of the most challenging problems that cleaners must solve. It is important to get as much information as possible from the customer about the source of the soil, odor or stain. If the source is known, it is much easier to choose the correct process to remove the problem.

Soils

Garments can become soiled from many different sources.

These include:

- Inorganic dust, dirt and particulates from the air
- *Inorganic* dirt from rubbing the garment on dirty surfaces
- *Organic* oils from soot and pollen

- Food and drink
- Blood and urine
- Greases, tars, adhesives or other thick, sticky substances
- Aged skin cells that have flaked off the body

With the exception of the organic oils and greases, most of these soils do not penetrate deeply into the fibers of the fabric. However, many of these soils may be tightly attached to the garment.

Soils can be hard to remove when:

- They have been ground into the fabric structure by a strong force
- They are attached to the fabric by *electrostatic bonding* with the fibers
- The ragged surfaces of the soil particles cling to the surface of the fabric

Soil can be removed by:

- Dry mechanical agitation such as beating or scrubbing
- Dry mechanical spinning
- Flushing with a water, *surfactant* or solvent solution

The fabricare professional can also combine these methods if a single method does not remove the soil.

Odors

There are several sources of objectionable odors. These include:

- Decomposing perspiration and body oils ground into the garment
- Organic contaminants from foods, solvents, greases or oils
- Bacteria or mites attracted to organic contaminants on the garment
- Particles of smoke and soot from cigarettes or fires
- Synthetic *volatile organic compounds*—such as fuels, paints, and solvents—that have been transferred to the garment

Odors typically suggest that some organic process is taking place on the garment.

Either a volatile compound is evaporating, or a food particle or body cell is decomposing, or bacteria are *metabolizing*. The odor itself is not on the garment, but the source of the odor is.

The techniques that can remove odor are:

- Removing the source of the odor by agitation, steaming or washing with a water, surfactant or solvent solution
- "Airing out" the garment over time in an oxygen rich atmosphere to allow the source of the odor to completely decompose
- Disinfecting the garment with a bleach or peroxide
- Heating the garment to promote rapid oxidation
- Masking the odor with a more dominant odor or perfume (this method does not remove the odor; it only covers up the odor temporarily)

The cleaner can also combine these methods if a single method does not remove the source of the odor.

Stains

Stains tend to colour a fiber in much the same way that a commercial dye does. Sources of stains include:

- Ink
- Dye
- Solvents
- Food and drink
- Perspiration
- Blood

Some stains are drawn up through the fibers by capillary action, the same way juice or coffee is soaked up by a paper towel. These stains stick to the surface of the garment and yarns, but do not penetrate the fibers. Other stains soak into the fibers themselves. Depending on the type of fiber, stains that penetrate into the chemical structure of a fiber may "set" over time. These stains can be very difficult to remove.

The more a stain contrasts with the colour of the garment, the more of a problem it is. A red ink stain on a white shirt must be completely removed, but a red ink stain on a red, floral garment may only need to be hidden. The stain can be hidden by reducing its colour so that it blends into the background.

Stains can be removed or hidden by:

- Flushing with a water, surfactant or solvent solution
- Bleaching or dyeing

The cleaner can also combine these methods if a single method does not treat the stain successfully.

SEPARATING SOILS, ODORS AND STAINS FROM FABRICS

Separating soil, odor sources and stains from fabric is a two-step process: "get it off" and "keep it off."

From the point of view of wet cleaning, this process consist of:

- Wetting a fabric and the soil, odor source or stain in order to separate the problem from the fabric
- Holding the soil, odor source or stain away from the fabric to prevent it from soiling the fabric again during the cleaning process

Different separation techniques are effective at separating different soils, odor sources and stains. Some of these techniques are mechanical processes; others involve the use of chemical agents.

Processes

Mechanical agitation removes dirt from a garment by either direct force or by breaking and pulverizing dirt particles. Breaking and pulverizing results in smaller particles, which are able to leave the fabric structure more easily.

These smaller particles are able to leave the fabric more easily than larger particles. Mechanical agitation does not work for all soils, however. For example, mechanical agitation alone may remove sand from a garment but may not be as effective at removing ground-in dirt. Other techniques are also required for removing stains.

Water is a powerful solvent that can help separate certain soils and stains from fabrics. Some soils dissolve in water that is used to clean a garment. The mechanical force of water flushing can also remove some soils, even though the soils themselves do not dissolve in water.

Adding certain chemicals to water can help water clean more effectively. These chemical additives decrease the surface tension of the water. This allows the water to penetrate the fabric more rapidly and thoroughly. These chemical additives are called "surfactants" because of their effects on surface tension.

Steaming and heating a garment also assist in separating soils and stains from a fabric. Heat can melt or soften particles. Steam can cause particles to expand or lose their electrostatic charges. Finally, drying can assist in soil separation as particles curl, break or change shape during the evaporation process. A dehydrated particle or cell is less likely to decompose rapidly or to attract bacteria and, therefore, less likely to generate odors.

Chemical Agents

Detergents. Detergents are a particular kind of surfactant. Detergents penetrate into tight spaces such as those between fibers and soils. Surfactants ease the process of separation by surrounding the soil particle and easing it away from the fabric and into the water. Once a surfactant has surrounded a soil particle and released it from a fabric, the problem is to keep the soil particle from attaching itself to the fabric again. Electrostatic charges help solve this problem.

Most surfactants and most fabrics carry negative electrostatic charges in water. These charges repel each other. Every time a soil particle surrounded by surfactant comes close to the fabric, the negative charges repel each other. This keeps the soil particles from reattaching themselves to the fabric. The soil particles stay suspended in water until they are flushed away in a rinse. Figure illustrates this process.

Surfactants (detergents) are made up of two different parts. One portion has an attraction to water which allows the detergent to be soluble in water. The other repels water but is attracted to fatty and oily compounds. When surfactants are dissolved in water that contains fabric with fat or oil spots, the surfactants are attracted to the fat or oil. Because surfactants are also attracted to water, they tend to pull the fat or oil away from the fabric and into the water. This helps remove these spots from fabrics.

Soaps: Soaps are a surfactant that is made when fatty acids are combined with alkali. Many soaps include extra alkali. This increases the effectiveness of the soap. Because most dirt is slightly acidic, the alkali in soaps tends to neutralize the dirt. This releases the chemical bonds that exist between the dirt and the fabric.

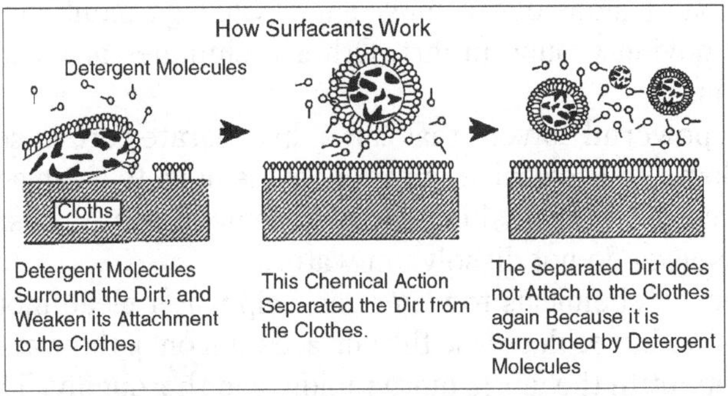

Synthetic Solvents: There are many synthetic surfactants and solvents on the market. Many of these solvents are made from petroleum. The most common synthetic solvent used for garment cleaning is perchloroethylene. Stoddard solvent, which is distilled from petroleum, is also commonly used in dry cleaning.

Bleaches and Oxidizers: Bleaching does not clean garments; it merely disinfects them. Whitening, odor removal and some organic stain removal is often accomplished by disinfection. Disinfection kills bacteria and mites that may accumulate on organic soils. Disinfection is accomplished by oxidation. Chlorinated bleaches oxidize and dissolve organisms that can then be flushed away in rinses. The speed of oxidation can be increased by exposing fabrics to certain forms of oxygen such as ozone or peroxide.

Digestors: The separation of some organic materials such as food or blood from a fabric can be accomplished by biological means. Enzymes are biologically active organisms that eat and digest organic soils such as food and blood. This loosens the soils and breaks them down into other materials that can be rinsed away.

Finishing Agents: Finishing agents are used for a number of purposes. For example, they restore body, resilience and smoothness to garments after processing. They also make fabric more soil resistant. This is accomplished by forming a film around the surface of the fiber. The film makes it more difficult for soil to become attached to the fiber. If soil does become attached to the fiber, the film also makes it easier to remove that soil.

GARMENT CLEANING METHODS

CURRENT AND ALTERNATIVE GARMENT CLEANING METHODS

Most of the commercial garment cleaning in the United States is currently done using the dry cleaning process. In the future, however, new cleaning processes may also be widely used. Some of these new processes are:

- Carbon dioxide
- Ozone
- Ultrasonic cleaning
- Wet cleaning

All of these processes have been or are currently being studied as methods for commercial garment cleaning. Only wet cleaning is currently ready to be used as an alternative to dry cleaning on a large commercial scale. The other processes require further testing and development.

DRY CLEANING

Conventional dry cleaning treats garments in either perchloroethylene or petroleum. Perchloroethylene is the solvent used in 80 per cent of dry cleaning plants. Petroleum is used in the remaining 20 per cent. The machinery used with these solvents has been modified over time resulting in four "generations" of machines. All four generations are still in use at this time. The first generation has separate washers and dryers, and operators transfer garments from the washer to the dryer. For this reason, these are called "transfer machines."

The second generation combined the washer and dryer into one unit. The third generation added controls to help reduce perchloroethylene vapor emissions. The fourth generation included additional controls that recycle the air in the machine in order to further reduce vapor emissions. As a result of these improvements, workers are exposed to less perchloroethylene vapor and the machines themselves use less perchloroethylene.

Dry cleaning has proven itself to be an efficient, affordable, costeffective process for cleaning a wide variety of fibers and fabrics without causing damage to garments. The primary drawback to dry cleaning is that perchloroethylene and petroleum solvents are hazardous substances. Because of the hazards, the use of these solvents is becoming increasingly regulated.

In order to comply with present and future regulations, the garment care industry has two options. First, the manufacturers can develop another generation of dry cleaning equipment with more effective controls. These machines would use less perchloroethylene and petroleum solvents, and release less of these solvents into the workplace and the environment. Second, the industry can invest in alternative cleaning processes that do not depend on hazardous solvents. In both cases, dry cleaning facilities will need to replace older equipment with new equipment.

Carbon Dioxide: Supercritical and Liquid

Research is being conducted using carbon dioxide (CO2) in both the *supercritical* and liquid state as a method of cleaning clothes. Carbon dioxide possesses several qualities that make it a desirable alternative for clothes cleaning:

- It is not *toxic* to humans
- It is easily recycled

- It is an inexpensive solvent
- It does not cause damage to *ozone* in the atmosphere
- It does not cause metal to corrode
- It is non-flammable
- It does not pollute ground water or soil
- It is readily available

Liquid carbon dioxide has become the primary focus of much of the research because it is safer to use and handle than supercritical carbon dioxide.

Carbon dioxide is a molecule formed naturally when two oxygen atoms join one carbon atom (C+O+O=CO2). Under normal circumstances, carbon dioxide is an *inert gas*. Carbon dioxide is familiar to everyone. When we breath in, our bodies absorb oxygen from the air. When we breath out, much of what we exhale is carbon dioxide gas. Carbon dioxide gas is typically found in very low concentrations in the atmosphere.

Under conditions of high temperature and pressure, CO2 reaches a state called the supercritical phase. Grease and oils dissolve when they come in contact with CO2 that is in the supercritical phase. This makes supercritical CO2 an effective cleaner. Supercritical CO2 has other advantages as well. When the pressure is reduced, grease and oil that are dissolved in the supercritical CO2 separate from the CO2. The grease and oil can then be disposed of, and the CO2 can be reused.

Supercritical CO2 has been used by the manufacturing industry to clean metal parts for some time. It has been tested in limited situations to determine if it can be used to clean garments. Further testing is required to determine if supercritical CO2 is effective, affordable, and safe to use as a garment cleaner.

Liquid CO2 exists at lower temperature and pressure than supercritical CO2. Liquid carbon dioxide was originally developed by the aerospace industry for cleaning optical and electrical parts.

It has been modified to dry clean garments. Liquid CO2 has no surface tension, which helps it separate dirt and grease from fabric. It can also be combined with surfactants to make it more effective. Like supercritical CO2, liquid CO2 can be recycled during the cleaning process. It has been reported that liquid CO2 can clean leather, furs and fine garments without causing damage or degradation.

Ozone Cleaning System

The Ozonated Water Continuous Cleaning System is a new process that is still being studied but is available commercially on a limited basis. The theory behind the process is that ozone (O3), working as an oxidizing agent, has the ability to disinfect, deodorize and bleach fabrics. Ozone can be added to water, in much the same way that carbon dioxide is used in carbonated soft drinks. Ozonated water can be combined with detergents to produce a garment that is cleaned, disinfected and deodorized.

The garment remains on a hanger throughout the wash, rinse, extraction and drying cycles. The ability to let the garment remain on the hanger during the cleaning process reduces the amount of labour used for finishing. The tendency of the garment to crease and wrinkle is decreased because each garment is processed individually. The company conducting the research suggests that with this process there is a reduced amount of labour required. The researchers report no shrinkage, change of shape, or wrinkling and claim that ornamental buttons do not need to be removed. They also state that it is cost effective to process clothes using this process.

Ultrasonic Cleaning

Recent research cooperatively funded by the U.S. Department of Energy and Kansas City Plant and Garment Care, Inc. determined that agitation for a continuous-flow, water-based clothes washing process could be provided by ultrasound. When clothes cleaned with ultrasound were examined, the clothes were clean enough to encourage researchers to continue developing this process.

Ultrasonic cleaning works by the pressure of ultrasonic waves directed at the fabric through a solution of water and soaps. Ultrasonic waves are sound waves at a frequency that is higher than the human ear can hear. Ultrasound works because molecules of soil are more dense than molecules of fiber. The energy from the ultrasound ex- cites the molecules of soil and causes them to separate from the fabric. Ultrasonic cleaning is currently being investigated as a possible alternative to conventional dry cleaning. The method has been shown to be effective, but more testing is required to determine the effect of ultrasound on various fabrics.

Wet Cleaning

Cleaning garments with water is an age-old technology which is currently attracting attention from consumers and cleaners interested in using an environmentally friendly method. Unlike traditional laundering methods, wet cleaning combines the gentle mechanical action of hand washing with the convenience of machine laundry. The success of wet cleaning is dependent upon more than just the machine itself. Wet cleaning machine operators need to be knowledgeable in the area of fiber science because wet cleaning requires a certain amount of observation and skill. Success is also dependant upon the combination of quality detergents, proper training and commitment to the process.

Wet cleaning uses a wet cleaning machine, detergents, additives, stain removal agents and a specialized dryer. Conventional additives to the wet cleaning process are not considered environmentally hazardous. The cleaning process consists of essentially the same steps as conventional dry cleaning.

Basically the process can be broken down into sorting, stain removal, wet cleaning, drying and finishing.

A COMPARISON OF DRY CLEANING AND WET CLEANING

Dry cleaning and wet cleaning are currently the two most efficient, affordable and cost-effective processes for cleaning garments. Figure illustrates the dry cleaning process. Figure illustrates the wet cleaning process. In general, these two cleaning processes are similar.

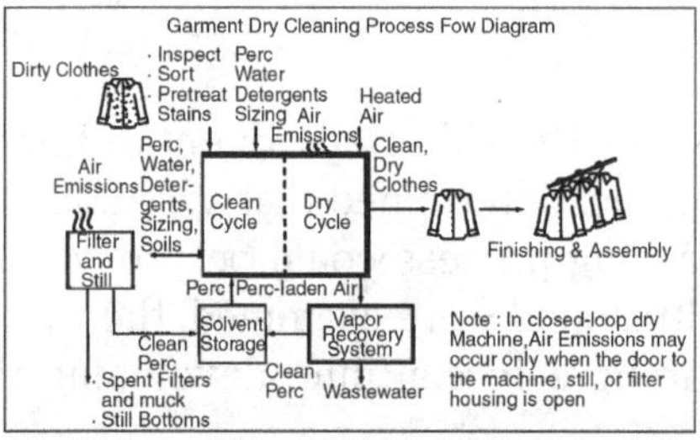

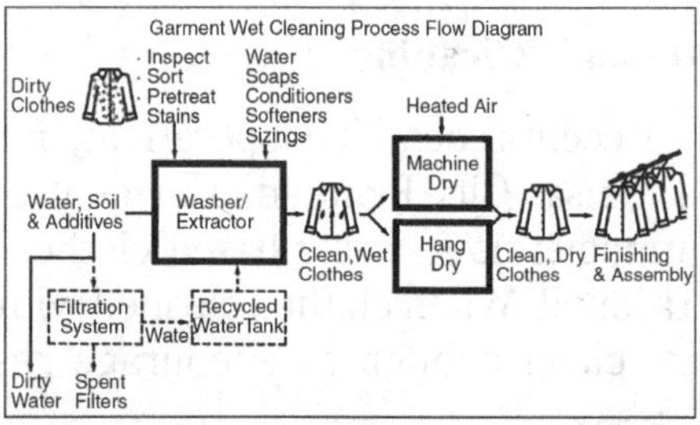

The basic steps in each process are preparation, washing, drying, and finishing and assembly.

Preparation

In both processes, the cleaner inspects, sorts and removes stains from the garments before washing. Inspection is basically the same for dry and wet cleaning. In both cases, the cleaner inspects the garments to determine what problems may occur in the cleaning process and to determine the best way to clean the garments.

If two dry cleaners were given the same garments and asked to sort them, they probably would not sort them the same way. If two wet cleaners were given the same test, they probably would not sort the garments the same way, either. Even so, the dry cleaners would be using one set of general rules for sorting, and the wet cleaners would be using another set of rules. The general rules for sorting are different because the cleaning processes are different.

The differences between cleaning with solvents and cleaning with water also result in differences in the stain removal process:

- Because the wet cleaning machine removes some water-based stains, these stains do not have to be removed before washing.
- Pretreated garments need to be dry before they can be put in a dry cleaning machine. Pretreated garments do not need to be dry before going into a wet cleaning machine. In fact, it is better in some cases to make sure that pretreated garments do not dry out, because this can cause rings to form on the fabric. This means that the wet cleaner can remove stains and immediately put garments in the wet cleaning machine.

- Several manufacturers have developed stain removers specifically for wet cleaning. Many of these products are less toxic than stain removing agents that are used for dry cleaning.

Washing and Drying

The washing and drying processes are the heart of the differences between dry cleaning and wet cleaning.

These differences include:

- Dry cleaning machines and wet cleaning machines have different controls.
- With the exception of older transfer machines, most dry cleaning machines both wash and dry. Wet cleaning machines only wash garments. Separate dryers are used for drying garments. Many of these dryers have been developed specifically for wet cleaning.
- Some wet cleaners choose to partially or completely air dry certain garments. This can require more space in which to hang the garments. It can also increase turnaround time.
- Somewhat more care is initially required to monitor the wet cleaning process in order to ensure that garments are cleaned successfully. The cleaner must have a good knowledge of fibers, fabrics, and the effects of the wet cleaning process.
- Dry cleaning machines use organic solvents like perchloroethylene or petroleum as the primary cleaning agent. Water, detergent and sizing are added to increase the efficiency of the dry cleaning process. Wet cleaning machines use water and detergent as the primary cleaning agent. Soap, sizing, conditioner and softener are added to the wet cleaning process.
- The waste products produced by dry cleaning machines include air and water that are hazardous. Special filters and control devices are required to minimize the amount of toxic waste that comes in contact with workers and the environment. These hazardous wastes also expose dry cleaners to liability if the waste is not disposed of properly. Wet cleaning machines do not produce hazardous waste, though some stain removal agents may be hazardous.

Finishing and Assembly

Garments that have been wet cleaned will require more work in the finishing step.

This happens most often when the cleaner is learning the wet cleaning process.

Less time will be required to finish these garments as the cleaner and the finisher gain more experience with the wet cleaning process. The purchase of state-of-the-art finishing equipment can help shorten finishing time.

HANDS-ON WET CLEANING

SORTING

Here it is covers general information on the sorting process.

The following additional steps will be covered at the sorting station:

- Sort loads just as to machine manufacturers' and detergent manufacturers' instructions
- Suit pieces should be sorted into same load
- Sort complete full loads and half loads
- Ties should be pinned inside pant leg or put in a mesh bag
- Put all knit sweaters in mesh bag

STAIN REMOVAL

Here we are covers general information on testing for colorfastness, and covers stain removal.

The following information and steps will be covered at the stain removal station:

- Start in the morning with dark loads, because dark loads tend to have fewer stains. This means that dark loads can be put in the washer more quickly. While the dark loads are being washed, the cleaner can remove stains from light-colored garments.
- Stain removal agents should be biodegradable and non-toxic. If dry cleaning solvents are used, waste should be collected and disposed of as hazardous waste.
- Excellent lighting will enhance results
- Test for colorfastness

WET CLEANING

The following information and steps will be covered at the wet cleaning station: n Match machine programme to load size and type n If using for both wet cleaning and laundry, switch machine to wet clean cycle n Use only EPA-approved soaps and detergents

- If the water is very soft, decrease detergent and increase sizing to protect fabric
- Watch for cracks in the door tubing that can cause leaking
- Take clothes out upon completion of cycle
- If using for both wet cleaning and laundry, rinse between cycles
- Clean drum after waterproofing or fire retardant cycle
- Clean suit pieces in the same load
- Process jackets in the afternoon so the jackets have time to hang overnight
- When cleaning down garments use a quick fill and extract cycle to completely wet the down

DRYING

The following information and steps will be covered at the drying station:

- Use a large enough dryer for adequate tumbling
- Use the shortest cycle possible to prevent relaxation damage
- Take clothes out immediately after processing
- Do not dry clothes to less than 10 per cent humidity
- Air dry garments when necessary
- Very fine and loose weave sweaters should be dried flat
- Molded black plastic form hangers work best

FINISHING

The following information and steps will be covered at the finishing station:

- Finish garments only when the amount of moisture in the garment is the same as the amount of moisture in the air of the cleaning facility
- Less finishing time is required if the finisher uses stretching equipment
- Conditioners, including starches, can reduce finishing time on some garments
- To bring back nap, brush suedes with a copper wire brush towards the nap
- Heavy wool coats should be brushed with a carding brush to pick up the pile
- Wipe velvet with a velvet brush to get all the pile going in the same direction

WET CLEANING EQUIPMENT

WET CLEANING MACHINES

Wet cleaning machines have been designed to process a wide range of garments, including garments that are usually dry cleaned. Manufacturers have focused on mechanical agitation and programmable controls to give the wet cleaner flexible options for safe, efficient cleaning.

Some of the new wet cleaning machines use a gentle, rolling action to minimize the risk of damage to fabric. These machines provide wash speeds as low as 5 to 10 rpm and programmable wash action that allows for minimal agitation, such as 1 second on and 60 seconds off. Other wet cleaning machines use air bubbles to provide gentle agitation. Gentle wet cleaning machines also have programmable spin speeds for extracting water from garments. Extraction cycles can be set to match the garment and fabric weight, so that the maximum amount of water is removed for each type of garment. Additional controls are available for water use, water level, temperature, heating, and automatic chemical injection. These features allow wet cleaners to precisely control the wet cleaning process. These controls also allow the wet cleaner to automatically process garments that would ordinarily be hand washed.

The drum capacity or load size of wet cleaning machines ranges from 10 to 135 pounds. An important fact to keep in mind is that the wet cleaning capacity of the machine is 50-60 per cent of the laundry capacity. This means that if the cleaner plans to routinely process 25 pound loads on a wet cleaning cycle, the machine purchased should have a minimum capacity of 50 pounds.

The amount of water used per wet cleaning cycle should be considered when choosing a wet cleaning machine, particularly in areas where water and sewer costs are high. Studies done in Chicago and Canada show that switching from dry to wet cleaning may increase water use. However, the studies also show that wet cleaning was still economically favorable to dry cleaning.

Machines with automatic detergent feeds should also have either an alarm on the detergent and/or additive tanks, or an automatic shut off. These features ensure that detergents and finishes are added when garments are processed. Processing garments without detergents and finishes can result in wasted labour, detergents and utility costs. The machine should also be equipped with enough detergent dispensers to allow the cleaner to process the full range of garments that he or she wishes. If skins are to be processed, the machine needs to be equipped with 2 dispensers in addition to the dispensers needed for normal garments on the wet clean cycle.

All currently available wet cleaning machines can be used as either a wet clean system or a standard laundry machine. This allows the cleaner to process shirts and "dry clean only" garments in one machine. Another advantage to purchasing a wet clean system is the ability to process skins. Leathers and furs are generally a high profit item for the professional cleaner and the ability to process them inhouse can increase profits.

DRYERS

A number of dryers have been developed specifically for wet cleaning. Most machines have programmable temperature and humidity controls that help the cleaner precisely set the amount of moisture that remains in the garment at the end of the drying process.

Even though the dryers have sophisticated temperature and humidity controls, the cleaner must pay careful attention to the load size and the mix of garments and fabric weights. Because different fabrics dry at different rates, sorting garments by fabric type and weight is even more important for drying than it is for wet cleaning. Cleaners can also programme drying cycles specifically for loads that contain a mix of fabrics, but care is needed to ensure that all garments dry properly.

Both electric and steam dryers use considerable energy. An electric dryer may need up to a 100 amp service to run properly. Cleaners who are considering installing a new dryer need to find out if they need to install 100 amp service to run it. One dryer manufacturer will be modifying their machine from a three phase to a single phase in the future, but for now the 100 amp service is necessary. If the cleaner is considering a steam dryer, he or she must determine if the current boiler has the capacity to support the dryer. A gas dryer should not be used in a shop that uses perc.

Though garments should be removed from the dryer when they are still slightly moist to prevent shrinkage, it is often necessary to let the garment dry completely before finishing. It is possible to let the garments air dry, or to purchase equipment to speed up this final drying process. Wind whips are available to blow dry garments, especially those with heavy areas like shoulder pads and thick waist bands. Drying cabinets can be used to dry garments that may be damaged by agitation of any sort.

WET CLEANING EQUIPMENT FEATURES AND SPECIFICATIONS

Wet cleaning technology is developing rapidly and machines available today may have different features. For a copy of the full report contact Environment Canada.

8

Dyeing

INTRODUCTION

Dyeing of textiles has been practiced for thousands of years with the first written record of the use of dyestuff dated at 2600 BC in China. All dyes were natural substances obtained from plant, animal or mineral sources. In 1856, William Henry Perkin, while searching for a cure for malaria, discovered the first synthetic dye, Mauve. The mauve dye was a brilliant fuchsia colour, but faded easily. Since that time, a great number of synthetic dyes have been manufactured and their resistance to running and fading has been almost eliminated. Almost all garments purchased today are dyes with synthetic dyes.

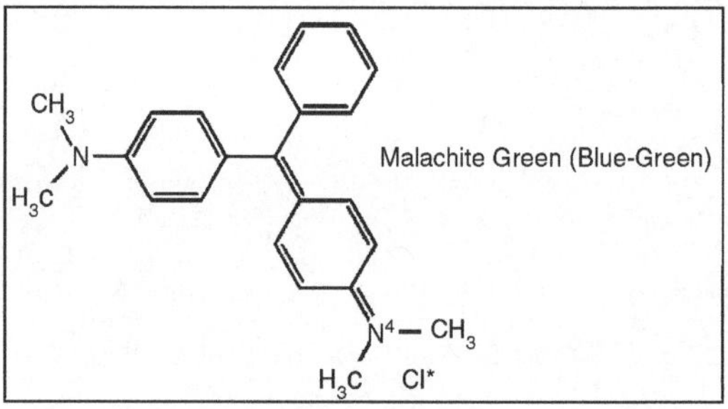

Dyes used for fabric such as cotton, wool, and silk are complex organic molecules that contain what is known as a chromophore group, that is, they contain some type of conjugated, alternating double and single bonds in part of the molecule. These molecules can absorb certain wavelengths of visible light and reflect the remaining light and, thus, give a fabric its colour. Not only do the dyes have polar or ionic groups, but fabrics such as cotton and wool also contain polar groups such as –OH (hydroxyl) and –NH (amide) which help the dye attach to the fabric. Sometimes chemical bonds are formed between the dye and the fabric molecules which hold the two together. Another process involves the use of *a mordant,* which serves as sort of an intermediary that bonds the dye and the fabric. If the dye molecules attach firmly to the fabric, the colour will be "fast", that is, it does not run when wet or washed, after the initial rinsing of excess dye. This experiment will deal with four types of dyeing processes, direct dyes, mordant dyes, developed dyes, and vat dyes, along with an example of how dyes can be used to identify different fabrics.

Direct Dyes are molecules that adhere to the fabric molecules without help from other chemicals. In this experiment as suggested, use a dye known as malachite green to illustrate this process.

Mordant Dyes are dyes that do not adhere to fabrics directly. These need a chemical intermediate, known as a mordant, to attach themselves to the fabric. In this process, the mordant is applied to the fabric and then the dye is applied, which then bonds to the mordant. In this experiment, as suggested, use tannic acid as a mordant to dye a sample of cotton with malachite green dye.

Developed Dyes are those where the dye-forming chemical reaction is carried out on the fabric. This process is commonly done with diazo dyes which are characterized by containing a –N=N– structure in the molecule. In this experiment, as suggested, use a dye known as para red which will be formed on the cotton fabric from the reaction between *p*-nitroaniline and 2-naphthol.

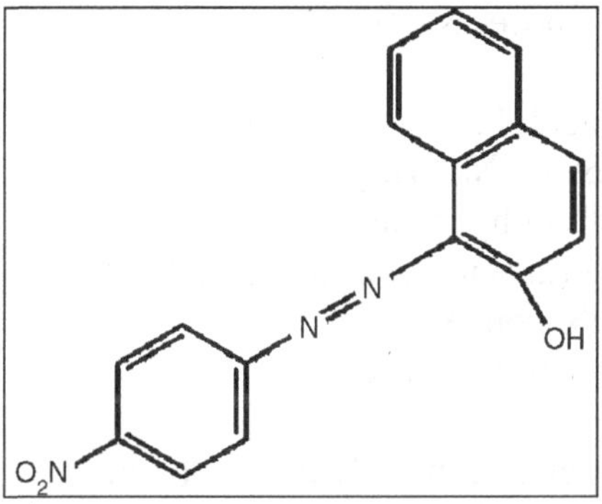

Vat Dyes are insoluble in their colored form. They are reduced by another chemical and converted to a soluble form. The reduced dye is applied to the fabric, and then exposed to the air which oxidizes he dye back to its colored form. In this experiment indigo dye will be used to dye a sample of cotton fabric. Indigo is the dye used to make blue jeans blue.

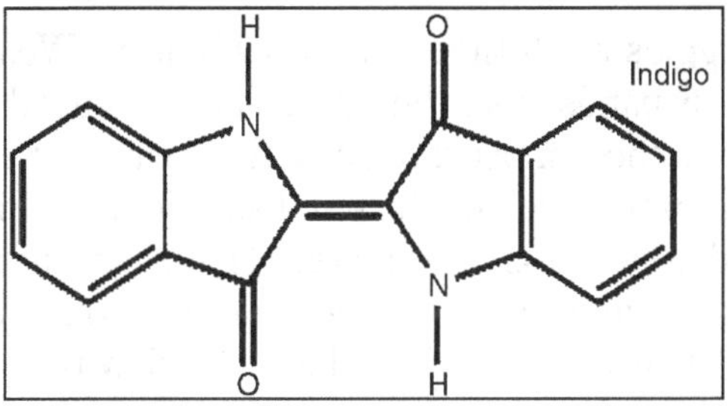

MATERIALS NEEDED

- 6 beakers, 400 mL
- Beaker, 20 mL or 50 mL

- Erlenmeyer flask, 125 mL, with stopper
- 3 stirring rods
- Graduated cylinders, 10 mL and 100 mL
- Tongs or forceps
- Beaker tongs
- Metal spatula
- Malachite green
- P-nitroaniline
- Tannic acid
- Sodium nitrite, NaNO2
- 2-naphthol
- Indigo powder
- TIS Identification stain No. 1
- TIS Identification stain No. 3A
- Iron(II) sulfate, FeSO4
- Sodium hydrosulfite, Na2S2O4
- .2H2O
- Hydrochloric acid, HCl, concentrated
- Sodium hydroxide, NaOH, pellets
- 5 per cent sodium hydroxide solution, NaOH
- 10 per cent acetic acid solution, HC2H3O2
- Ammonia, NH3, concentrated
- Thermometer, -10 to 120°C
- Ice
- Cotton, strips, approximately 2 cm x 5 cm
- Wool, strips, approximately 2 cm x 5 cm
- silk, strips, approximately 2 cm x 5 cm
- multifiber fabric, strips, approximately 2 cm x 5 cm
- Rubber or plastic gloves

SAFETY PRECAUTIONS

Wear safety goggles at all times in the laboratory Wear rubber or plastic gloved to protect your hands from the dyes. The dyes used in this experiment should be handled with the utmost care since they will dye the skin and clothing.

Be careful to keep the dye powders and solutions off the desk tops. Clean up any spills immediately. Most dye materials are irritants to the skin, eyes, and respiratory system and may be toxic by inhalation and ingestion. In the event of skin contact, wash the affected area well with water. Hydrochloric acid is corrosive and can cause severe burns to the skin along with irritating fumes.

Work with the concentrated solution under a hood. In the event of skin contact, wash the affected area well with water. If the skin is broken or blistered, seek medical assistance. If irritated by the vapors, get fresh air immediately.

Sodium hydroxide is caustic and can cause severe burns to the skin. Do not handle sodium hydroxide pellets with your hands.

In the event of skin contact, wash the affected area well with water until there is no soapy feel. If the skin is broken or blistered, seek medical assistance. Sodium nitrite is an irritant to the skin, eyes, and respiratory system.

It is toxic by ingestion. In the event of skin contact, wash the affected area well with water. Sodium nitrite can also be an explosion hazard. Do not heat this compound dry or in contact with any flammable material.

Ammonia vapors are irritants to the eyes and respiratory system. Work with the concentrated solution under a hood. In the event of skin contact, wash the affected area well with water. If irritated by the vapors, get fresh air immediately.

DISPOSAL AND CLEAN UP

Dispose of all materials in the specific containers provided in the laboratory. Dye stains can be removed from glassware by applying a paste of powdered bleach and water, then rinsing.

The paste will also remove stains from the hands, which should be washed off with soap and water.

PROCEDURE

Direct Dyeing

Measure 0.1 g of malachite green. Dissolve the malachite green in 200 ml of distilled or deionized water in a 400-mL beaker. Heat the solution to boiling and place one strip each of cotton, wool, silk, and multifiber fabric in the solution. Allow the cloth strips to remain in the hot solution for 2 to 3 minutes.

Remove the strips of cloth with forceps or tongs and rinse thoroughly with water. Are the colors "fast"? Record your results. Save the solution for the next dyeing procedure.

Mordant Dyeing

Measure 0.2 g of tannic acid. Dissolve the tannic acid in 100 ml of distilled or deionized water in a 400-mL beaker.

Heat the solution to boiling and place a strip of cotton and multifiber fabric in the solution. Allow the cloth strips to remain in the hot solution for 1 to 2 minutes.

Remove the cloth strips from the tannic acid solution, place them on a paper towel and press the excess solution from them, then immerse in the hot malachite green solution saved from the first procedure.

Allow the cloth strips to remain in the hot malachite green solution for 2 minutes. Remove the cloth strips with forceps or tongs and rinse thoroughly with water. Are the colors "fast"? Record your results.

Developed Dyeing

Measure 3 mL of concentrated hydrochloric acid. Add it to a 400-mL beaker. Place the beaker under a fume hood. Measure 0.1 g of p-nitroaniline. Add it to the hydrochloric acid in the 400-mL beaker. Slowly, with stirring, add 10 mL of distilled or deionized water to the mixture. Place the beaker on an ice bath and cool the solution to about 5°C. Measure 0.1 g of sodium nitrite and dissolve it in 5 ml of distilled or deionized water in a small beaker. Add this solution to the cold p-nitroaniline solution. Measure 0.1 g of 2-naphthol (?-naphthol). Add it to 5 ml of 5 per cent sodium hydroxide solution. Immerse a strip of cotton and multifiber fabric in the naphthol solution. Remove the cloth strips and press to remove excess solution. Now immerse the cloth strips in the cold solution previously prepared. After the dye (para red) forms, test for colorfast properties and record your observations.

VAT DYEING

Measure 0.1 g indigo powder in a 125-mL Erlenmeyer flask. Add 0.1 g of sodium hydrosulfite (Na2S2O4), 2 pellets of sodium hydroxide, and 10 mL of water to the flask. Stopper and shake for 2 to 3 minutes. Dilute to about 50 ml with distilled or deionized water. Holding a cotton strip with forceps, dip the cloth into the dye solution for about 20 seconds. Remove the cloth and expose to air for a few minutes. Record your observations Holding a multifiber fabric strip with forceps, dip the cloth into the dye solution for about 20 seconds. Remove the cloth and expose to air for a few minutes. Record your observations. Test both cloth strips for colour fastness and record the results.

ALTERNATIVE PROCEDURE

If sodium hydrosulfite is not available, a mixture of 0.2 g indigo, 1.0 g ferrous sulfate, 5 ml water, and 5 ml concentrated ammonium hydroxide can be used. Heat the solution to boiling for a few minutes, allow the ferric hydroxide to precipitate, pour off the solution.

IDENTIFICATION OF AN UNKNOWN FABRIC

Dyeing tests involve the use of T.I.S. Identification stains. These are special mixtures of dyes that colour different fibers different colors used to identify an unknown fabric or fibers.

Fabric Identifaction using T.I.S. Stain No. 1

Solution to boiling. Maintain a hot, but not actively boiling solution. Wet samples of cotton, wool, silk, and multifiber fabric, with distilled or deionied water. Squeeze out the excess liquid and place the samples in the hot dye bath for 3 to 5 minutes. Remove the samples and wash out any excess dye. Compare the colour of the cloth samples with the multifiber fabric.

Fabric Identifaction using T.I.S. Stain No. 3A

A second colour test can be conducted using T.I.S. Identification Stain No. 3A. Prepare a 0.05 per cent solution of T.I.S. Identification Stain No. 3A. (0.05 g for each 100 mL water) Heat the solution to boiling and add 1 mL 10 per cent acetic acid solution. Maintain a hot, but not actively boiling solution.

Wet samples of cotton, wool, silk, and multifiber fabric, with distilled or deionied water. Squeeze out the excess liquid and place the samples in the hot dye bath for 5 minutes. Remove the samples and wash out any excess dye. Compare the colour of the cloth samples with the multifiber fabric.

NATURAL DYEING OF TEXTILES

Dyeing is an ancient art which predates written records. It was practised during the Bronze age in Europe. Primitive dyeing techniques included sticking plants to fabric or rubbing crushed pigments into cloth. The methods became more sophisticated with time and techniques using natural dyes from crushed fruits, berries and other plants, which were boiled into the fabric and gave light and water fastness (resistance), were developed.

Fig. Indigo.

Some of the well known ancient dyes include madder, a red dye made from the roots of the Rubia tinctorum, blue indigo from the leaves of Indigofera tinctoria, yellow from the stigmas of the saffron plant, and dogwood, an extract of pulp of the dogwood tree. The first use of the blue dye, woad, beloved by the Ancient Britons, may have originated in Palestine where it was found growing wild. The most famous and highly prized colour through the age was Tyrian purple, noted in the Bible, a dye obtained from the spiny dye-murex shellfish. The Phoenicians prepared it until the seventh century, when Arab conquerors destroyed their dyeing installations in the Levant. A bright red called cochineal was obtained from an insect native to Mexico. All these produced high-quality dark colours. Until the mid-19th century all dyestuffs were made from natural materials, mainly vegetable and animal matter.

Today, dyeing is a complex, specialised science. Nearly all dyestuffs are now produced from synthetic compounds. This means that costs have been greatly reduced and certain application and wear characteristics have been greatly enhanced. But many practitioners of the craft of natural dying (*i.e.* using naturally occurring sources of dye) maintain that natural dyes have a far superior aesthetic quality which is much more pleasing to the eye. On the other hand, many commercial practitioners feel that natural dyes are non-viable on grounds of both quality and economics. In the West, natural dyeing is now practised only as a handcraft, synthetic dyes being used in all commercial applications. Some craft spinners, weavers, and knitters use natural dyes as a particular feature of their work.

In many of the world's developing countries, however, natural dyes can offer not only a rich and varied source of dyestuff, but also the possibility of an income through sustainable harvest and sale of these dye plants. Many dyes are available from tree waste or can be easily grown in market gardens. In areas where synthetic dyes, mordants (fixatives) and other additives are imported and therefore relatively expensive, natural dyes can offer an attractive alternative.

The knowledge required for sourcing and extracting such dyes and mordants is, however, often not available as extensive research work is required to identify suitable plants, minerals, etc. In Zambia for example, there is a wealth of plants available for producing

natural dyes, but due to lack of knowledge of the processes involved in harvesting and processing the plants, little use is made of this natural resource. In some countries, such as India, Nigeria and Liberia, where this research has been carried out, or where there exists a tradition of natural dyeing, natural dyes and mordants are used widely.

TYPES OF TEXTILES SUITABLE FOR DYING

Natural dyes can be used on most types of material or fibre but the level of success in terms of fastness and clarity of colour varies considerably. Users of natural dyes, however, tend to also use natural fibres, and so as suggested, look in more detail at this group. Natural fibres come mainly from two distinct origins, animal origin or vegetable origin. Fibres from an animal origin include wool, silk, mohair and alpaca, as well as some others which are less well known. All animal fibres are based on proteins. Natural dyes have a strong affinity to fibres of animal origin, especially wool, silk and mohair and the results with these fibres are usually good. Fibres of plant origin include cotton, flax or linen, ramie, jute, hemp and many others. Plant fibres have cellulose as their basic ingredient. Natural dyeing of certain plant based textiles can be less successful than their animal equivalent. Different mordanting techniques are called for with each category. When a blend of fibre of both animal and plant origin is being dyed, then a recipe should be chosen which will accentuate the fibre which is required to be dominant.

EQUIPMENT NEEDED FOR HOME DYEING AND VERY SMALL-SCALE COMMERCIAL DYEING

Most equipment needed for dyeing fabrics at home, or at the very small-scale commercial level, can be found in almost any market place throughout the world.

The following is a list of the equipment requirements and a brief explanation of their use:

- *Heat source*: This can be any type of cooking stove; gas, wood, kerosene, charcoal, electricity. This is used for heating the liquid used during mordanting and dyeing.
- *Pestle and mortar*: Used for milling the natural dye or minerals, where this is called for.
- *Mordanting and dyeing pans*: Stainless steel or enamel pans are the most suitable for dyeing. The size of pan depends upon the quantities of fabric that will be dyed. Do not use pans made from copper, aluminium or iron, unless absolutely necessary, as these metals have properties which can change the colour of the dye.
- *Tirring rods*: Stainless steel or glass rods are best as they can be cleaned and used for different colour dyes. If wooden stirring rods are used then there should be a different spoon for each colour.
- *Thermometer*: This is used to measure the temperature of the liquid during mordanting and dyeing. A long thermometer (to reach the liquid at the bottom of the pan) is preferred, with a range of 0 – 100oC (32 – 210oF).
- *Measuring jugs*: These are used to measure the quantities of liquid called for in the recipe. Sometimes precise quantities are called for.
- *Storage containers*: Used for storing the dyestuffs and mordants. Large glass and plastic jars are ideal. Some mordants and dyes are sensitive to light and should therefore be stored in sealed light-proof containers.
- *Plastic bowls and buckets*: A variety of plastic bowls or buckets of varying sizes are useful when wetting or rinsing fabrics.
- *Strainer*: Used for straining the liquid off the dyestuff in the dyebath.
- *Weighing scales*: Used for obtaining the correct quantities as specified in the recipe. A scales with metric and imperial measurement is useful as conversions from one system to the other are not then needed.

- *Protective equipment*: Gloves for holding hot pans will prevent burns. An apron will protect your clothing. Rubber gloves will prevent skin irritation caused by mordants, and will also prevent you from dyeing your hands. A face mask can cut down the amount of fumes or powder inhaled during the dyeing process.

MORDANTS

Few natural dyes are colour-fast with fibres. Mordants are substances which are used to fix a dye to the fibres. They also improve the take-up quality of the fabric and help improve colour and light-fastness. The term is derived from the Latin *mordere,* to bite. Some natural dyes, indigo for example, will fix without the aid of a mordant; these dyes are known as 'substantive dyes'. Others dyes, such as madder and weld, have a limited fastness and the colour will fade with washing and exposure to light Traditionally, mordants were found in nature. Wood ash or stale urine may have been used as an alkali mordant, and acids could be found in acidic fruits or rhubarb leaves (which contain oxalic acid), for example. Nowadays most natural dyers use chemical mordants such as alum, copper sulphate, iron or chrome (there are concerns, however about the toxic nature of chrome and some practitioners recommend that it is not used). Mordants are prepared in solution, often with the addition of an 'assistant' which improves the fixing of the mordant to the yarn or fibre. The most commonly used mordant is alum, which is usually used with cream of tartar as an additive or assistant. Other mordants are:

- Iron (ferrous sulphate)
- Tin (stannous chloride)
- Chrome (bichromate of potash)
- Copper sulphate
- Tannic acid
- Oxalic acid

Using a different mordant with the same dyestuff can produce different shades, for example;

- *Iron* is used as a 'saddener' and is used to darken colours.
- *Copper sulphate* also darkens but can give shades which are otherwise very difficult to obtain.
- *Tin* brightens colours.
- *Tannic acid*, used traditionally with other mordants, will add brilliancy.
- *Chrome* is good for obtaining yellows.
- *Oxalic acid* is good for extracting blues from berries.
- *Cream of Tartar* is not really a mordant but is used to give a lustre to wool.

Mordants are often poisonous, and in the dye-house they should be kept on a high shelf out of the reach of children. Always use protective clothing when working with mordants and avoid breathing the fumes.

The mordant can be added before, during or after the dyeing stage, although most recipes call for mordanting to take place prior to dyeing. It is best to follow the instructions given in the recipe being used or experiment on a sample before carrying out the final dyeing. Later in this brief as suggested, explain how the mordant is mixed and used as part of the dyeing process.

These chemical mordants are usually obtained from specialist suppliers or from chemists. Where this is prohibitive, due to location or cost, natural mordants can be used. There are a number of plants and minerals which will yield a suitable mordant, but their availability will be dependent upon your surroundings.

- Some plants, such as mosses and tea, contain a small amount of aluminium. This can be used as a substitute to alum. It is difficult to know, however, how much aluminium will be present and experimentation may be necessary.
- Iron water can be used as a substitute to ferrous sulphate. This can be made simply by adding some rusty nails and a cupful of vinegar to a bucket-full of water and allowing the mixture to sit for a couple of weeks.
- Oak galls or sumach leaves can be used a substitute to tannic acid.
- Rhubarb leaves contain oxalic acid.

NATURAL DYESTUFFS

Dyestuffs and dyeing are as old as textiles themselves. Nature provides a wealth of plants which will yield their colour for the purpose of dyeing, many having been used since antiquity.

As suggested, look at some of these naturally occurring dyes, their source and the colours they produce. Later in the brief as suggested, look at the application of the dyes to textiles.

Fig. Marigold.

Almost any organic material will produce a colour when boiled in a dye-bath, but only certain plants will yield a colour that will act as a dye. The plants given in Table are a selection of plants that have stood the test of time, and are used widely and traditionally by natural dyers.

Natural dyes fall into the following categories:

- Leaves and stems
- Twigs and prunings
- Flower heads
- Barks
- Roots
- Outer skins, hulls and husks
- Heartwoods and wood shavings
- Berries and seeds
- Lichens
- Insect dyes

The choice of mordant for a particular plant is dependant upon the material with which it will be used. It is necessary to check a recipe before using a plant, or one can experiment to see what effect a mordant has for a particular application.

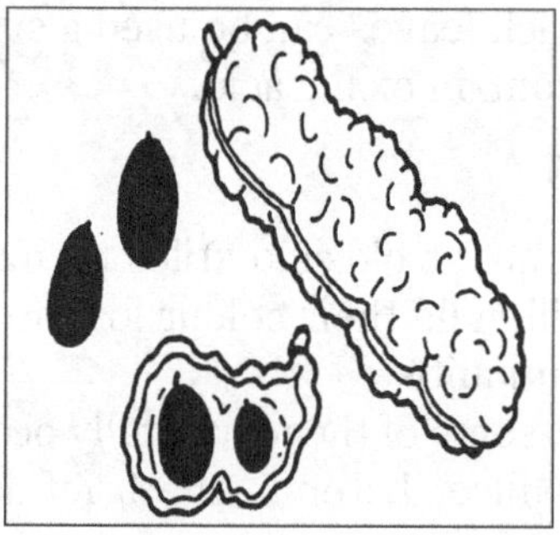

Fig. Groundnuts.

It is recommended that plants be grown specifically for the purpose of dyeing. Harvesting plants from the wild on a non-sustainable basis can endanger the survival of the plant. Many lichens are registered as protected organisms and it is illegal to gather them from the wild. One source of lichen is from timber mills where logged trees are being processed, but ask first!

TESTING DYES

It is always useful and interesting to test the dye which is to be used on a sample of the yarn or fabric to be dyed. The outcome will depend on the fabric, the mordant that has been used and dye that has have been chosen. Testing is best carried out on a series of marked (for identification) samples, which have been mordanted with a number of different mordants. Tests can be carried out for light, water and washing fastness using simple standard test methods. Box 1 shows a simple test for washing fastness. Similar tests exists for water and light fastness.

Processing of the plant can take one of many forms, but usually takes the form of soaking or boiling the plant to extract the dye. Some plants, such as indigo, need special preparation for use.

Some plants will need boiling to extract their dye while others can merely be soaked for extended periods.

PROCESS OF DYEING

Application of the Dye

Dyeing can be carried out at any of the following stages in the textile manufacturing stage:

- The fibres can be dyed before they are spun. *Fibre dyeing* provides a deep penetration of the dye into the fibre, giving even colour and excellent colour-fastness.
- The yarn can be dyed after spinning but before the product is woven or otherwise fabricated. This is called *package dyeing.*

Before the fabric is finished, it can be dyed in lengths (*piece dyeing*).This process allows manufacturers the opportunity to produce fabrics in their natural colours, and then dye them to order.

- In *cross-dyeing*, fabrics of two or more fibres can be dyed so that each fibre accepts a different dyestuff and becomes a different colour, through the use of appropriate dyestuffs for each fibre.

It is essential for the correct identification of the fibre or other fabric to be made before dyeing commences.

METHODS OF DYEING

There are a number of methods of applying dye to a fabric. Although the most common method used for applying natural dyes is the vat method, there are techniques which have been developed to allow patterns to be incorporated during the dying process.

It is worth bearing in mind that using natural dyes is a complex art and the skills required for using natural dyes are learned over many years. Don't be put off if you don't get the desired results at the first attempt!.

Vat Dyeing

In the simplest form of dyeing a textile material is immersed in dye and gradually brought to the boil. Alternatively the fibre is allowed to sit and soak for several hours or days. During this period, agitation is necessary to allow full penetration of the textile by the dyestuff. Depending on the type of fabric and dyestuff used, certain salts or acids may be added to assist absorption of the dye.

The principal difficulty in dyeing mixed yarns and fabrics is to achieve the same colour in both fibres. Cotton fibres may, for instance, absorb dyes rapidly, while the wool fibres will have to be boiled over an extended period to reach the same depth of shade. This could lead to significant damage to the material. In this case a chemical compound would need to be used to restrain the rate at which the cotton fibre takes up the dyestuff.

The amount of dyestuff which is used is usually given in the recipe. It is usually quoted as a percentage weight of the fabric to be dyed. Box 2 demonstrates a typical, simple recipe for dyeing using natural dyestuff, in this case onion skins.

Batik

Batik is a starch resist-dyeing process, developed on the Island of Java in modern-day Indonesia. Colour is prevented from reaching certain areas of a fabric by covering these areas with molten wax. The fabric is starched prior to the design being drawn upon it. The wax is applied with a type of cup with a fine pouring spout, usually made of copper. The technique has been developed to a high art form in Indonesia from where it is exported to many parts of the world. Batik paintings, as well as sarongs and lengths of fabric, are produced. When the fabric is dyed, all waxed areas resist the dyestuff. The wax is then removed by placing the fabric in boiling water. For patterns with many colours the same procedure is repeated until the full design is completed.

Tie-dyeing

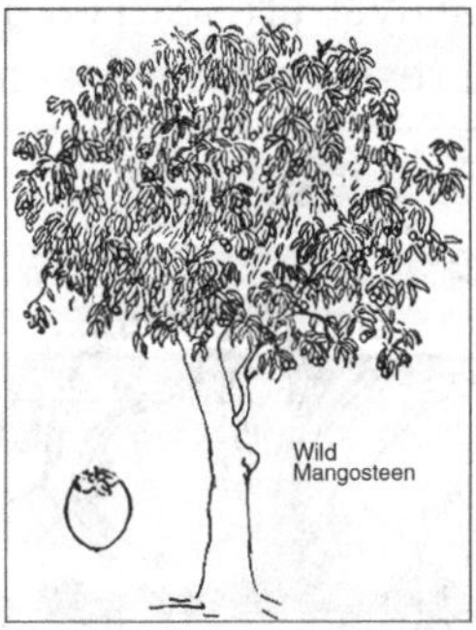

Tie-dyeing is another popular artisanal dyeing technique. In this resist-dyeing process, waxed thread is tightly tied around the areas chosen to resist the coloured dyestuff, and the fabric is dipped into the dye. The waxed thread is then removed and the fabric dried. This process can be repeated for each colour to be added.

OBTAINING UNUSUAL COLOURS

Colours other than those obtained by simply using a single dye, can be obtained by mixing dyes or dyeing a fabric more than once in different dyebaths. The colour triangle is a useful tool in determining the colours needed to produce a required shade.

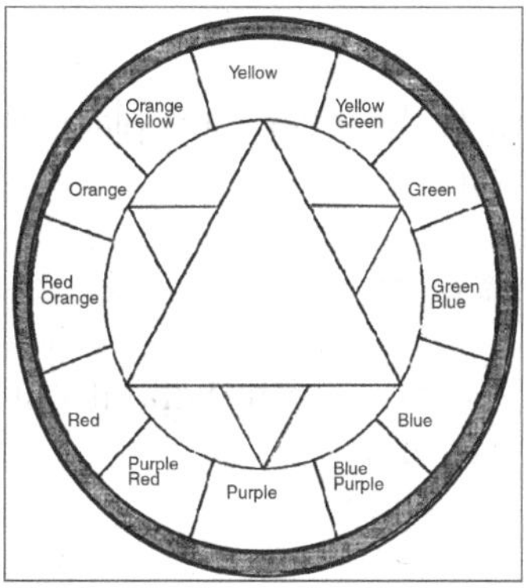

FABRIC DYEING TIPS

METHODS

- *Washing Machine*: Most convenient.
- *Stovetop*: For very dark colors like black, where the amount of fabric will fit in a large pot.
- *Sink or Bucket*: Suitable for small or delicate items being dyed a light colour.

WASHING MACHINE METHOD

STEP 1

- Take 1 package or a ½ bottle of RIT dye (2 packs or a full bottle will give a full colour particularly for medium, bright or dark colors) and place it in an old sauce pan with a quart of water (1 pint of water per package used).
- Place sauce pan on the stove and heat it up, stirring while it heats. The purpose here is to totally dissolve the dye into the water.

STEP 2

- Fill your washer with hot water (140 degrees is best for dark colors), add the dye/water from your saucepan, plus a splash of vinegar to help set the dye, and turn on the agitator for several minutes- to mix in the dye solution.

STEP 3

- Take a piece of rag and put it in the washer for a few minutes. Pull it out and look at the colour to see if it's what you want. If its too light then add more dissolved dye.
- Never add more than 4 packages or 2 bottles per washer load when dyeing a large amount of fabric.

STEP 4

- Turn machine on again, and let it agitate for another 3 minutes. — This will assure that all the dye is thoroughly mixed in.
- Never pour dye directly on fabric.

STEP 5

- Wet the garments to be dyed and place them loosely into washer and turn machine on.

STEP 6

- Take garments out of machine after the cycle and either line dry or put in dryer.
- If you use the dryer, be sure to wipe it out with a bleach cloth after use to remove any dye.

STEP 7

Clean-Up:

- Fill washer to highest water level.
- Add a small amount of detergent plus a cup of bleach.
- Run a complete cycle.
- Finish by wiping the top of the tub and inside lid with bleach.
- Clean out lint trap.
- Plastic or rubber machine parts may be tinted but won't stain future laundry loads.
- Clean up saucepan and any other items used.

9

Sewing

AN OVERVIEW

Sewing or stitching or tailoring is the fastening of cloth, leather, furs, bark, or other flexible materials, using needle and thread. Its use is nearly universal among human populations and dates back to Paleolithic times (30,000 BCE). Sewing predates the weaving of cloth.

Sewing is used primarily to produce clothing and household furnishings such as curtains, bedclothes, upholstery, and table linens. It is also used for sails, bellows, skin boats, banners, and other items shaped out of flexible materials such as canvas and leather.

Most sewing in the industrial world is done by machines. Pieces of a garment are often first tacked together. The machine has a complex set of gears and arms that pierces thread through the layers of the cloth and semi-securely interlocks the thread.

Some people sew clothes for themselves and their families. More often home sewers sew to repair clothes, such as mending a torn seam or replacing a loose button. A person who sews for a living is known as a seamstress (from seams-mistress) or seamster (from seams-master), dressmaker, tailor, garment worker, or machinist.

"Plain" sewing is done for functional reasons: making or mending clothing or household linens. "Fancy" sewing is primarily decorative, including techniques such as shirring, smocking, embroidery, or quilting.

Sewing is the foundation for many needle arts and crafts, such as applique, canvas work, and patchwork.

While sewing is sometimes seen as a semi-skill job, flat sheets of fabric with holes and slits cut into the fabric can curve and fold in complex ways that require a high level of skill and experience to manipulate into a smooth, ripple-free design. Aligning and orienting patterns printed or woven into the fabric further complicates the design process. Once a clothing designer with these skills has created the initial product, the fabric can then be cut using templates and sewn by manual laborers or machines.

SEAM ALLOWANCE

Seam allowance is the area between the edge of the fabric and the line of stitching. It is usually 1.5 cm away from the edge of the fabric except for the hem, where the seam allowance is usually 2.5 cm or more. This is often the case for standard home dressmaking. Industry seam allowances vary but are usually 0.6 cm.

OCCUPATIONS REQUIRING SEWING

- Bookbinding/Bookbinder
- Shoemaking/Cobbler
- Corsetier
- Draper
- Dressmaker
- Glove/Glover
- Hatter
- Milliner
- Parachute rigger
- Quilting/Quilter
- Sailmaker
- Seamstress
- Tailor
- Taxidermist
- Upholsterer
- Leatherworker

SEWING TOOLS AND ACCESSORIES

Stitching Awl

The Stitch-It-Awl is a hand sewing awl for safely and quickly sewing by hand mainly medium to heavy materials such as sailcloth, canvas, leather, soft luggage, upholstery or webbing. It was developed to solve the problems associated with the use of existing hand sewing devices for heavy materials, such as the needle and leather palm used by sail makers.

Bobbin

A bobbin is the little plastic or metal wheel that fits underneath the needle plate of a sewing machine. Thread runs off the bobbin, gets picked up by the needle's thread and forms the bottom part of a seam.

The bobbin is an essential part of a sewing machine. It enables the machine to form a seam from two threads, which makes a stronger, smoother, neater seam. Even the best hand sewer cannot duplicate the look of a machine seam. Bobbins, in one form or another, have been around for years. They have always been a part of the sewing machine, even in the old treadle machine days. Using the term bobbin for this implement probably dates from the days of weaving, since the weft threads of a loom are wound around bobbins. The word bobbin is of indeterminate age and origin, which in English, means it probably dates back at least to Middle English.

The sewing machine bobbin was wound by hand for early machines, but the advent of the electric motor sewing machine meant that the bobbin could be wound on the machine itself. After all, the motor turned, so why not put it to another useful purpose?

Depending on the sewing machine, a bobbin may be wound on top of the machine or underneath the needle plate. Winding a bobbin on top of the sewing machine is not difficult, but the operator must make sure to wind the thread in the right direction. Putting the bobbin on upside down will wind the thread off the bobbin.

Some sewing machines require the user to put the bobbin underneath the needle plate, hold the needle thread and turn the hand-wheel until the needle thread "catches" the bobbin thread. Other machines have a "drop-in" bobbin, so the user only needs to make certain the bobbin is oriented correctly and drop it into the correct place. The machine then picks up the bobbin thread on its own.

A sewer should make sure that the sewing machine's bobbin is of the proper size for the machine. A plastic bobbin should not have cracks or splits. Bobbins are available for most sewing machine models in stores where sewing machines and fabrics are sold, as well as online.

Bodkin

Several different implements go by the name of bodkin, including a weapon, an archaic type of hairpin, and two different sewing notions. As a weapon, a bodkin is a dagger; as a hairpin, it's an ornament with a stiletto shape; and as a sewing tool, it's either an instrument similar to an awl for piercing holes in cloth or a long needle with a large eye and blunt tip for drawing tape, elastic, or ribbon through a casing.

The needle-like bodkin is also an item of interest. It was the tool used to lace stays and bodices. A bodkin was stored in a bodkin case, which looks rather like a flat rectangular version of a lipstick container.

Modern bodkins may have several special features. Some needle-like bodkins have a ballpoint tip for ease of movement through the casing. A flat bodkin is a small metal tool that resembles a matchstick with a hole (or holes) in the end for threading. A third type of bodkin is a piece of wire with one end covered with a plastic tip and the other formed into a loop. A completely different style of bodkin is fashioned like a miniature tongs with gripping teeth on its ends to grasp the material being threaded and a neck ring to tighten and hold it in place.

It may prove useful to know that in the absence of a threading bodkin, several other tools can be substituted. A tapestry needle, with its large eye and blunt tip, can serve the purpose of a bodkin by tying the material around the eye so it can't come loose. A large safety pin can also be used by piercing the material to be drawn through the casing, closing the pin, and working it around in the same manner as the bodkin.

Dress Form

A dress form is used to give a three-dimensional view on the substance of clothing that is being sewed. They come in all sizes and shapes for almost every substance of clothing that can be made. When a piece of clothing is made, it can be put on the dress form so one can see how the piece of clothing will turn out. Then one can make alterations upon the clothing after seeing what it looks like on a body.

Scissors

Scissors are hand-operated cutting instruments. They consist of a pair of metal blades, or tangs, connected in such a way that the sharpened edges slide against each other. Scissors are used for cutting various thin materials, such as paper, cardboard, metal foil, thin plastic, cloth, rope and wire. Scissors can also be used to cut hair and food.

Scissors and shears exist in a wide variety of forms depending on their intended uses. Children's scissors, used only on paper, have dull blades and rounded corners to ensure safety. Scissors used to cut hair or fabric must be much sharper. The largest shears used to cut metal or to trim shrubs must have very strong, sharp blades.

Specialized scissors include sewing scissors, which often have one sharp point and one blunt point for intricate cutting of fabric, and nail scissors, which have curved blades for cutting fingernails and toenails.

Special kinds of shears include pinking shears, which have notched blades that cut cloth to give it a wavy edge, and thinning shears, which have teeth that cut every second hair strand, rather than every strand giving the illusion of thinner hair.

Tape Measure

A tape measure or measuring tape is a flexible form of ruler. It consists of a ribbon of cloth, plastic, fiber glass, or metal strip with linear-measurement markings. It is a common measuring tool. Its flexibility allows for a measure of great length to be easily carried in pocket or toolkit and permits one to measure around curves or corners.

Today it is ubiquitous, even appearing in miniature form as a keychain fob, or novelty item. Surveyors use tape measures in lengths of over 100 m (300+ ft).

Sewing Needle

A sewing needle is a long slender tool with a pointed tip. The first needles were made of bone or wood; modern ones are manufactured from high carbon steel wire, nickel- or gold plated for corrosion resistance. The highest quality embroidery needles are plated with two-thirds platinum and one-thirds titanium alloy. Traditionally, needles have been kept in needle books or needle cases which have become an object of adornment.

A needle for hand sewing has a hole, called the eye, at the non-pointed end to carry thread or cord through the fabric after the pointed end pierces it. Hand sewing needles have different names depending on their purpose.

Needle size is denoted by a number on the packet. The convention for sizing is that the length and thickness of a needle increases as the size number decreases. For example, a size 1 needle will be thicker and longer, while a size 10 will be shorter and finer.

Pattern (Sewing)

In sewing and fashion design, a pattern is an original garment from which other garments of a similar style are copied, or the paper or cardboard templates from which the parts of a garment are traced onto fabric before cutting out and assembling (sometimes called paper patterns).

Patternmaking, pattern making or pattern cutting is the art of designing patterns. A custom-fitted basic pattern from which patterns for many different styles can be created is called a sloper or block.

Patterns for Custom Dressmaking

A custom dressmaker frequently employs one of three pattern creation methods. The flat-pattern method begins with the creation of a sloper or block, a basic pattern for a fitted, jewel-neck bodice and narrow skirt, made to the wearer's measurements. The sloper is usually made of lightweight cardboard or tagboard, without seam allowances or style details. Once the shape of the sloper has been refined by making a series of mock-up garments called toiles (UK) or muslins (US), the final sloper can be used in turn to create patterns for many styles of garments with varying necklines, sleeves, dart placements, and so on.

Although it is also used for women's' clothing, the drafting method is more commonly employed in men's garments and involves drafting a pattern directly onto pattern paper using a variety of straightedges and curves.

The pattern draping method is used for more elaborate and unique designs that are hard to obtain through the flat pattern method. This is because it is nearly impossible to account for the way a fabric will drape or hang on the body without an actual 3-dimensional test run. It involves creating a muslin mock-up pattern by pinning fabric directly on a dress form, then transferring the muslin outline and markings onto a paper pattern or using the muslin as the pattern itself.

Patterns for Home Sewing

Home sewing patterns are generally printed on tissue paper and sold in packets containing sewing instructions and suggestions for fabric and trim. Modern patterns are available in a wide range of prices, sizes, styles, and sewing skill levels, to meet the needs of consumers.

Home sewing patterns are graded, that is, redrawn to fit larger and smaller sizes than the original design. Ebenezer Butterick invented the commercially produced graded home sewing pattern in 1863 (based on grading systems used by Victorian tailors), originally selling hand-drawn patterns for men's and boys' clothing. In 1866, Butterick added patterns for women's clothing, which remains the heart of the home sewing pattern market today.

There are some applications today that enable a home sewer to customize a computerized pattern to fit her body measurements and or body shape. The 3D technology enables the home sewer to see a virtual simulation of the final garment as it will appear on her. This reduces the Time-to-Market as well as the number of muslins/test garments that are needed.

A variation on the theme was evolved by iconic British brand Clothkits. Clothkits devised ingenious cut and sew clothing kits for home sewing that avoided the need for paper patters. Rather than using conventional techniques, Clothkits pre printed fabric with both quirky designs and the pattern lines, to make dressmaking for the novice easier.

Pin

A pin is a device used for fastening objects or material together. It is usually made of steel, or on occasion copper or brass. It is formed by drawing out a thin wire, sharpening the tip, and adding a head. Nails are related, but are typically larger.

Pincushion

A pincushion (or less commonly pin cushion) is a small cushion, typically 3-5 cm across, which is used in sewing to store pins or needles with their heads protruding so as to take hold of them easily, collect them, and keep them tidy.

Pincushions are typically filled tightly with stuffing, so as to hold pins rigidly once placed. The stereotypical design – a tomato with a strawberry attached – seems to have been around forever but, most likely, was introduced during the Victorian Era.Typically, the tomato is filled with wool roving to prevent rust, and the strawberry is filled with an abrasive to clean the pins and keep them sharp.

Pincushions come in all shapes, sizes and forms. Magnetic pincushions have the advantage of being able to easily keep the work area clear of pins, but the drawback that the pins become magnetized easily and an attempt to retrieve one may pick up several.

Rotary Cutter

A rotary cutter is a tool generally used by quilters to cut fabric. It consists of a handle with a circular blade that rotates, thus the tool's name. Rotary blades can be found in different sizes: usually smaller blades are used to cut around corners, while larger blades are used to cut to straight lines. Several layers of fabric can be cut simultaneously, making it easier to cut out patchwork pieces of the same shape and size than with regular scissors. In conjunction with a rotary cutting ruler – specially designed to be used with a rotary cutter, and made of thick and resistant plastic – it is also a practical tool for achieving the straight, squared lines that are so important in patchwork. The first rotary cutter was introduced by the Olfa company in 1979.

Seam Ripper

A seam ripper, stitch ripper or stitch unpicker is a small tool used for unpicking stitches.

The most common form consists of a handle, shaft and head. The head is usually forked with one side of the fork flattening out and becoming a blade and the other side forming a small point. In some designs the blade side then tapers back to a point to allow easier insertion in tight stitching.

In use the blade is inserted into the seam underneath the thread to be cut. The thread is allowed to slip down into the fork and the tool is then lifted upwards allowing the blade to cut through the thread. Once the seam has been undone in this way the loose ends can be removed and the seam resewn.

Sewing Table

A sewing table or work table is a table or desk used for sewing. Generally is has large amounts of space and a full set of sewing tools. Nearby there will be a chair and a waste bin. A common attachment is a dropleaf to give expanded space. Other attachments can be a cloth bag for storing sewing materials, drawers, or shelves.

Chalk

Chalk is a soft, white, porous sedimentary rock, a form of limestone composed of the mineral calcite. Calcite is calcium carbonate or $CaCO_3$. It forms under relatively deep marine conditions from the gradual accumulation of minute calcite plates (coccoliths) shed from micro-organisms called coccolithophores. It is common to find chert or flint nodules embedded in chalk. Chalk can also refer to other compounds including magnesium silicate and calcium sulfate.

Chalk is resistant to weathering and slumping compared to the clays with which it is usually associated, thus forming tall steep cliffs where chalk ridges meet the sea. Chalk hills, known as chalk downland, usually form where bands of chalk reach the surface at an angle, so forming a scarp slope. Because chalk is porous it can hold a large volume of ground water, providing a natural reservoir that releases water slowly through dry seasons.

Thimble

A thimble is a protective shield worn on the finger or thumb. It is generally used for sewing.

The earliest known thimble was Roman and was found at Pompeii. Made of bronze, its creation has been dated to the first century AD. A second Roman thimble was found at Verulamium, present day St Albans, in the UK and can be viewed in the museum there.

United Kingdom Detector Finds Database, thimbles dating to the tenth century have been found in England, and thimbles were in widespread use there by the 1300s. Although there are isolated examples of thimbles made of precious metals — Elizabeth I is said to have given one of her ladies-in-waiting a thimble set with precious stones — the vast majority of metal thimbles were made of brass. Medieval thimbles were either cast brass or made from hammered sheet. Early centers of thimble production were those places known for brass-working, starting with Nuremburg in the 15th century, and moving to Holland by the 17th.

In 1693, a Dutch thimble manufacturer named John Lofting established a thimble manufactory in Islington, in London, England, expanding British thimble production to new heights. He later moved his mill to Buckinghamshire to take advantage of water-powered production, resulting in a capacity to produce more than two million thimbles per year. By the end of the 18th century, thimble making had moved to Birmingham, and shifted to the "deep drawing" method of manufacture, which alternated hammering of sheet metals with annealing, and produced a thinner-skinned thimble with a taller shape. At the same time, cheaper sources of silver from the Americas made silver thimbles a popular item for the first time.

Thimbles are usually made from metal, leather, rubber, and wood, and even glass or china. Early thimbles were sometimes made from whale bone, horn, or ivory. Natural sources were also utilized such as Connemara marble, bog oak, or mother of pearl. Rarer works from thimble makers utilized diamonds, sapphires, or rubies.

Advanced thimblemakers enhanced thimbles with semi-precious stones to decorating the apex or along the outer rim. Cabochon adornments are sometimes made of cinnabar, agate, moonstone, or amber. Thimble artists would also utilize enameling, or the Guilloché techniques advanced by Peter Carl Fabergé.

Originally, thimbles were used solely for pushing a needle through fabric or leather as it was being sewn. Since then, however, they have gained many other uses. In the 1800s they were used to measure spirits, which brought rise to the phrase "just a thimbleful". Prostitutes used them in the practice of thimble-knocking where they would tap on a window to announce their presence. Thimble-knocking also refers to the practice of Victorian schoolmistresses who would tap on the heads of unruly pupils with dames thimbles.

Before the 18th century the small dimples on the outside of a thimble were made by hand punching, but in the middle of that century, a machine was invented to do the job. If one finds a thimble with an irregular pattern of dimples, it was likely made before the 1850s. Another consequence of the mechanization of thimble production is that the shape and the thickness of the metal changed. Early thimbles tend to be quite thick and to have a pronounced dome on the top. The metal on later ones is thinner and the top is flatter.

Collecting thimbles became popular in the UK when many companies made special thimbles to commemorate the Great Exhibition held in the Crystal Palace in Hyde Park, London.

In the 19th century, many thimbles were made from silver; however, it was found that silver is too soft a metal and can be easily punctured by most needles. Charles Horner solved the problem by creating thimbles consisting of a steel core covered inside and out by silver, so that they retained their aesthetics but were now more practical and durable. He called his thimble the Dorcas, and these are now popular with collectors. There is a small display of his work in Bankfield Museum, Halifax, England.

Early American thimbles made of whale bone or tooth featuring miniature scrimshaw designs are considered valuable collectibles. Such rare thimbles are prominently featured in a number of New England Whaling Museums.

During the First World War, silver thimbles were collected from "those who had nothing to give" by the British government and melted down to buy hospital equipment. In the 1930s and 40s red-topped thimbles were used for advertising. Leaving a sandalwood thimble in a fabric store was a common practice for keeping moths away.

Thimbles have also been used as love-tokens and to commemorate important events. A miniature thimble is one of the tokens in the game of Monopoly. People who collect thimbles are known as digitabulists.

Yarn

Yarn is a long continuous length of interlocked fibres, suitable for use in the production of textiles, sewing, crocheting, knitting, weaving, embroidery and ropemaking.

Thread is a type of yarn intended for sewing by hand or machine. Modern manufactured sewing threads may be finished with wax or other lubricants to withstand the stresses involved in sewing. Embroidery threads are yarns specifically designed for hand or machine embroidery.

Tracing Paper

Tracing paper is a type of translucent paper. It is made by immersing uncut and unloaded paper of good quality in sulphuric acid for a few seconds. The acid converts some of the cellulose into amyloid form having a gelatinous and impermeable character. When the treated paper is thoroughly washed and dried, the resultant product is much stronger than the original paper. Tracing paper is resistant to oil grease and to a large extent impervious to water and gas.

Tracing paper is named as such for its ability for an artist to trace an image onto it. When tracing paper is placed onto a picture, the picture is easily viewable through the tracing paper. Thus, it becomes easy for the artist to find edges in the picture and trace the image onto the tracing paper. Pure Cellulose fiber is translucent, and it is the air trapped between fibers, that makes paper opaque and looks white. If the fibers are refined and beaten until all the air is taken out, then the resulting sheet will be translucent. Translucent papers are dense and contain up to 10 per cent moisture at 50 per cent humidity. This type of paper is roughly 25 per cent lighter than regular paper.

The sizing in production will determine whether it is for laser printer or inkjet/offset printing. Tracing paper may be uncoated or coated. Natural tracing paper for laser printing is usually uncoated.

Tracing Wheel

A tracing wheel is an instrument with serrated teeth on a wheel attached to a handle. Used to transfer markings from patterns onto fabric with or without tracing paper, this sewing tool also makes slotted perforations. Such markings might include pleats, darts, buttonholes, notches or placement wikt:lines for appliques or pockets. There are two basic types of tracing wheels available to the modern sewer, one with a serrated edge and one with a smooth edge.

Wax

Traditionally, wax (or beeswax) is a substance secreted by bees and used in constructing their honeycombs.

The term has come to refer more generally to a class of substances with properties similar to beeswax:

- Plastic (malleable) at normal ambient temperatures
- A melting point above approximately 45 °C (113 °F) (which differentiates waxes from fats and oils)
- A relatively low viscosity when melted (unlike many plastics)
- Insoluble in water
- Hydrophobic

Waxes may be natural secretions of plants or animals, artificially produced by purification from natural petroleum or completely synthetic. In addition to beeswax, carnauba (a plant epicuticular wax) and paraffin (a petroleum wax) are commonly encountered waxes which occur naturally. Earwax is an oily substance found in the human ear.

SEWING EQUIPMENT

BAR TACK

Bar tack is a series of hand or machine made stitches used for reinforcing areas of stress on a garment, such as pocket openings, bottom of a fly opening or buttonholes. It consists of a series of close-set zig-zag stitches (machine) or whip-stitches (hand), usually 1/16"-1/8" in width and 1/4"-3/8" in length. In denim jeans, it is often in a contrasting colour, such as orange or white.

BLANKET STITCH

The blanket stitch is a stitch used to reinforce the edge of thick materials. Depending on circumstances, it may also be called a whip stitch or a crochet stitch. It is defined as "A decorative stitch used to finish an unhemmed blanket. The stitch can be seen on both sides of the blanket."

This stitch has long been both an application by hand and as a machine sewn stitch. When done by hand, it is sometimes considered a crochet stitch. When done by machine, it is called a whip stitch or, sometimes, a Merrow Machine Company Crochet Stitch, after the first sewing machine that was used to sew a blanket stich. This machine was produced and patented by the Merrow Machine Company in 1877.

The defining characteristic of the crochet machine is its ability to sew with yarn and stitch thick goods with a consistent overlock edge. From 1877-1925 the machine evolved dramatically, and consequently so did the capacity of manufacturers to produce goods with the whip stitch. To this day, Merrow retains a patent on the mechanism that creates the blanket stitch, renewing it with each redesign, which has occurred hundreds of times since its introduction in 1877. The most recent patent renewal occurred in 2007.

In the early 1990s Chinese manufacturers copied the design of the original crochet sewing machines; Merrow is currently involved in litigation with several Chinese firms who are accused of violating Merrow's intellectual property rights.

BUTTONHOLE STITCH

Buttonhole stitch and the related blanket stitch are hand-sewing stitches used in tailoring, embroidery, and needle lace-making.

Buttonhole stitches catch a loop of the thread on the surface of the fabric and needle is returned to the back of the fabric at a right angle to the original start of the thread. The finished stitch in some ways resembles a letter "L" depending on the spacing of the stitches. For buttonholes the stitches are tightly packed together and for blanket edges they are more spaced out. The properties of this stitch make it ideal for preventing raveling of woven fabric. Buttonhole stitches are structurally similar to featherstitches.

Applications

In addition to reinforcing buttonholes and preventing cut fabric from raveling, buttonhole stitches are used to make stems in crewel embroidery, to make sewn eyelets, to attach applique to ground fabric, and as couching stitches. Buttonhole stitch scallops, usually raised or padded by rows of straight or chain stitches, were a popular edging in the 19th century.

Buttonhole stitches are also used in cutwork, including Broderie Anglaise, and form the basis for many forms of needlelace.

Variants

Examples of buttonhole or blanket stitches include:

- Blanket stitch
- Buttonhole stitch
- Closed buttonhole stitch, in which the tops of the stitch touch to form triangles
- Crossed buttonhole stitch, in which the tops of the stitch cross
- Detached buttonhole stitch, in which rows of buttonhole stitches are worked to form a "floating" filling stitch
- Buttonhole shading, in which rows of buttonhole stitch are sewn in related colors to give a naturalistic shaded effect
- *Buttonhole stitches combined with knots include*:
  - Top knotted buttonhole stitch
  - German knotted buttonhole stitch
  - Tailor's buttonhole stitch
  - Armenian edging stitch

Buttonhole bars are parallel rows of thread laid across an open space in lace or cutwork and then completely covered with closely space buttonhole stitches.

CHAIN STITCH

Chain stitch is a sewing and embroidery technique in which a series of looped stitches form a chain-like pattern. Chain stitch is an ancient craft - examples of surviving Chinese chain stitch embroidery worked in silk thread have been dated to the Warring States period (5th-3rd century BC). Handmade chain stitch embroidery does not require that the needle pass through more than one layer of fabric. For this reason the stitch is an effective surface embellishment near seams on finished fabric. Because chain stitches can form flowing, curved lines, they are used in many surface embroidery styles that mimic "drawing" in thread. Chain stitches are also used in making tambour lace, needlelace, macramé and crochet.

CROSS-STITCH

Cross-stitch is a popular form of counted-thread embroidery in which X-shaped stitches in a tiled, raster-like pattern are used to form a picture. Cross-stitch is usually executed on easily countable evenweave fabric called aida cloth. The stitcher counts the threads in each direction so that the stitches are of uniform size and appearance. This form of cross-stitch is also called counted cross-stitch in order to distinguish it from other forms of cross-stitch. Sometimes cross-stitch is done on designs printed on the fabric (stamped cross-stitch); the stitcher simply stitches over the printed pattern.

DARNING

Darning is a sewing technique for repairing holes or worn areas in fabric or knitting using needle and thread alone. It is often done by hand, but it is also possible to darn with a sewing machine. Hand darning employs the darning stitch, a simple running stitch in which the thread is "woven" in rows along the grain of the fabric, with the stitcher reversing direction at the end of each row, and then filling in the framework thus created, as if weaving. Darning is a traditional method for repairing fabric damage or holes that do not run along a seam, and where patching is impractical or would create discomfort for the wearer, such as on the heel of a sock. Darning also refers to any of several needlework techniques that are worked using darning stitches:

- Pattern darning is a type of embroidery that uses parallel rows of straight stitches of different lengths to create a geometric design.
- Net darning, also called filet lace, is a 19th century technique using stitching on a mesh foundation fabric to imitate lace.
- Needle weaving is a drawn thread work embroidery technique that involves darning patterns into barelaid warp or weft threads.

LOCKSTITCH

A lockstitch is the most common mechanical stitch made by a sewing machine.

The lockstitch uses two threads, an upper and a lower. The upper thread runs from a spool kept on a spindle on top of or next to the machine, through a tension mechanism, a take-up arm, and finally through the hole in the needle. The lower thread is wound onto a bobbin, which is inserted into a case in the lower section of the machine. To make one stitch, the machine lowers the threaded needle through the cloth into the bobbin area, where a hook catches the upper thread at the point just after it goes through the needle. The hook mechanism carries the upper thread entirely around the bobbin case, so that it has made one wrap of the bobbin thread. Then the take-up arm pulls the excess upper thread (from the bobbin area) back to the top forming the lockstitch ideally in the center of the thickness of the material, the tension mechanism prevents the thread from being pulled from the spool side, the needle is pulled out of the cloth, and the feed dogs pull the cloth back one stitch length, the cycle is repeated as the machine turns mechanically.

Lockstitch is so named because the two threads, upper and lower, "lock" together in the hole in the fabric which they pass through. The term "single needle stitching", often found on dress shirt labels, refers to lockstitch, as opposed to chain stitch which unravels easily and is usually used on lower quality garments.

OVERLOCK

An overlock stitch sews over the edge of one or two pieces of cloth for edging, hemming or seaming. Usually an overlock sewing machine will cut the edges of the cloth as they are fed through (such machines are called 'sergers'), though some are made without cutters. The inclusion of automated cutters allows overlock machines to create finished seams easily and quickly. An overlock sewing machine differs from a lockstitch sewing machine in that it uses loopers fed by multiple thread cones rather than a bobbin. Loopers serve to create thread loops that pass from the needle thread to the edges of the fabric so that the edges of the fabric are contained within the seam. Overlock sewing machines usually run at high speeds, from 1000 to 9000 rpm, and most are used in industry for edging, hemming and seaming a variety of fabrics and products. Overlock stitches are extremely versatile, as they can be used for decoration, reinforcement, or construction.

Overlocking is also referred to as "overedging", "merrowing" or "serging". Though "serging" technically refers to overlocking with cutters, in practice the four terms are used interchangeably.

PAD STITCH

Pad stitches are types of running stitches, made by taking small stitches perpendicular to the line of stitching. Pad stitches secure two or more layers of fabric together and give the layers more firmness; smaller/denser stitches create more firmness. Pad stitches may also enforce an overall curvature of the layers. Tailors pad stitch a jacket's lapel and undercollar to give them more firmness and to help them maintain their curvature. The line of stitching usually runs parallel the direction of the most important curve of the layers. For example, pad stitches in a suit's lapel run parallel to the lapel's roll line; pad stitches in the undercollar of a tailored jacket run parallel to the collar back edge.

RUNNING STITCH

The running stitch or straight stitch is the basic stitch in hand-sewing and embroidery, on which all other forms of sewing are based. The stitch is worked by passing the needle in and out of the fabric. Running stitches may be of varying length, but typically more thread is visible on the top of the sewing than on the underside.

Running stitches are used in hand-sewing and tailoring to sew basic seams, in hand patchwork to assemble pieces, and in quilting to hold the fabric layers and batting or wadding in place. Loosely spaced rows of short running stitches are used to support padded satin stitch.

Running stitches are a component of many traditional embroidery styles, including kantha of India and Bangladesh, and Japanese sashiko quilting.

TACK (SEWING)

In sewing, to tack or baste is to make quick, temporary stitching intended to be removed.

Tacking is used in a variety of ways:

- To temporarily hold a seam or trim in place until it can be permanently sewn, usually with a long running stitch made by hand or machine called a tacking stitch or basting stitch.
- To temporarily attach a lace collar, ruffles, or other trim to clothing so that the attached substance may be removed easily for cleaning or to be worn with a different garment. For this purpose, tacking stitches are sewn by hand in such a way that they are almost invisible from the outside of the garment.
- To transfer pattern markings to fabric, or to otherwise mark the point where two pieces of fabric are to be joined. A special loose looped stitch used for this purpose is called a tack or tailor's tack.
- A basting stitch is essentially a straight stitch, sewn with long stitches and unfinished ends. The basting stitch is used for temporarily holding sandwiched pieces of fabric in place. The stitch is removed after the piece is finished. Often used in quilting or embroidery.

TOPSTITCH

Topstitching is a sewing technique. It is used most often on garment edges such as necklines and hems, where it helps facings to stay in place and gives a crisp edge. Decorative topstitching is designed to show, and may be done in a fancy thread or with a special type of stitch. Otherwise, topstitching is generally done using a straight stitch with a thread that matches the fashion fabric.

Bibliography

Ainley, Albert: *Woolen and worsted loomfixing; a book for loomfixers and all who are interested in the production of plain and fancy worsteds and woolens*, Lawrence Mass, 1900.

Alford, Marianne Margaret Compton Cust: *Needlework as art*, London: Sampson Low Marston Searle and Rivington, 1886.

Alfred Jenks & Son: *Illustrated Catalogue of Machines Built by Alfred Jenks & Son, Manufacturers of Every Variety of Cotton and Wool Carding, Spinning, and Weaving Machinery, in All Its Departments*, Bridesburg, Pa.: Alfred Jenks & Son, 1853.

Allen, Frederick James: *The shoe industry*, Boston: The Vocation bureau of Boston, 1916.

Allen, Nellie B: *Cotton and other useful fibers*, Boston: Ginn, 1929.

Allington, Sara May: *Practical sewing and dressmaking*, Boston: Page Co., 1913.

Allinson, May: *Dressmaking as a trade for women in Massachusetts*, Washington: Govt. print. off., 1916.

Atkinson, Edward: *Cotton and Cotton Manufactures in the United States*, Boston: Franklin Press, 1880.

Baird, Robert H: *The American Cotton Spinner, and Managers' and Carders' Guide*: A Practical Treatise on Cotton Spinning, Philadelphia: H. C. Baird, 1869.

Batchelder, Samuel: *Introduction and Early Progress of the Cotton Manufacture in the United States*, Boston: Little, Brown and Company, 1863.

Brown Brothers & Co: Brown Brothers & Co., Successors to Butler, Brown & Co.: *Manufacturers of Shaw's U. S. Standard Ring Travelers, Belt Hooks, Wire Goods, Etc., and Dealers in Supplies for Cotton, Woolen, Silk, Jute & Flax Mills, Providence*, R.I.: Brown Brothers & Co., c1881.

E. Jenckes Manufacturing Co: *Illustrated Price List: E. Jenckes Mfg. Co., Sole Manufacturers of Hicks' U. S. Standard Ring Travelers*, Manufacturers' Supplies of Every Description, Pawtucket, R.I.: E. Jenckes Mfg. Co., 1883.

Lowell Machine Shop: *The Lowell Machine Shop, Lowell, Mass.: Builders of Cotton Machinery of Every Description*, Lowell, Mass.: Stone, Bacheller & Livingston, 1882.

Mason Machine Works: *The Mason Machine Works, Taunton Massachusetts, U.S.A.: Inventors and Builders of Cotton Machinery*, Taunton, Mass.: The Company, 1898.

Merrill, Gilbert R: *American Cotton Handbook: A Practical Reference Book for the Entire Cotton Industry*, New York: American Cotton Handbook Co., c1941.

Scott, Robert: *The Practical Cotton Spinner, and Manufacturer: The Managers', Overlookers' and Mechanics' Companion*, Corr. and enl. Philadelphia: H. C. Baird, 1851.

Webber, Samuel: *Manual of Power for Machines, Shafts, and Belts, with the History of Cotton Manufacture in the United States*, New York: D. Appleton and Company, 1879

Index

A

Air-conditioning 5, 6, 7
Appropriate 183, 186, 188, 191
Automatic Control 4

B

Basic Clothing 15, 16

C

Capability Assumptions 99
Concept of Strategy 28
Conclusions 17, 94, 160
Conjunction 191
Conservation Technologies 1
Control Charts 81, 91, 93, 95, 96, 97, 98, 100
Core Competencies 31, 35, 36
Critical Evaluation 41, 46, 51, 58, 64, 66, 70, 73

D

Developing Core Competencies 32
Digital Printing 164
Drying Efficiency 11
Dyeing and Finishing 1, 8, 9, 15, 17, 141

E

Electric Heating 2
Electric Motor 2, 33, 259
Element Strategies 37
Energy Consumption 1, 5, 7, 8, 13, 14, 15, 17, 18
External environment 38, 40, 49, 50, 51, 52, 67

F

Fashion Clothing 15, 16
Fashion Design 172

G

Garment Industry 177

H

Heat exchanger 4, 7

I

Inlet Water 10, 11
Intervention Budget 104, 109

J

Jute Diversification 173, 174

K

Knitting 7, 14, 179, 180, 185, 193, 194, 265, 269

L

Lighting 1, 2, 238

M

Mathematical Definition 87
Multimedia Design 171

O

Objectives 19, 20, 21, 22, 23, 24, 27, 28, 29, 30, 31, 37, 38, 110, 182, 183
Organic Solvent Processing 12

P

Pie Chart 77, 78, 79, 80
Piping 3
Process Capability 93, 98, 99, 100
Process Printing Technique 165
Processing technique 9, 10
Production 164, 172, 174, 175, 176

R

Rapid Dyeing 10
Recycling of Drain 4
Ring Spinning Operation 6
Rise Margin 10
Run Chart 80, 81, 91

S

Solvent Processing 12
Sophisticated 183, 192
Space Saving 5, 13
Spot Printing Technique 166
Spraying Techniques 164
Steam Accumulators 4
Strategy 16, 25, 28, 29, 30, 31, 34, 35, 36, 37, 38, 39, 40, 48, 55, 57, 59, 60, 61, 67, 68, 69, 70, 71, 74, 119, 176, 177, 186
Subsequent 191, 192, 193

T

Technical Fields 1, 10, 17
Tendency 167
Textile Costing 115
Textile Costs 101, 102, 103, 105, 107, 108, 114, 115, 116
Textile industry 1, 2, 3, 4, 5, 13, 14, 16, 17
Textile Markets 15
Textile Printing 166

W

Weaving 1, 7, 14, 15, 176, 177, 179, 185, 192, 218, 229, 266, 268, 274